Cellulose Science and Derivatives

Edited by Arpit Sand and Sangita Banga

Published in London, United Kingdom

IntechOpen

Supporting open minds since 2005

Cellulose Science and Derivatives
http://dx.doi.org/10.5772/intechopen.94799
Edited by Arpit Sand and Sangita Banga

Contributors
Lethiwe D. Mthembu, Rishi Gupta, Nirmala Deenadayalu, Kaman Singh, Ashok Kumar, Satya Prakash Gupta, Agata Kołodziejczyk, Ramiro Escudero, González Jeovani, Pérez Rosa Elena, Cho Cho, Thinzar Aye, Aung Khaing, Takaomi Kobayashi, Soufian El Barkany, Issam Jilal, Zahra Bahari, Youssef El Ouardi, Hassan Amhamdi, Mohamed Loutoua, Mohamed Abou-Salama, Abderahmane El Idrissi, Amin Salhi, Katri Laatikainen, Nozieana Khairuddin, Mohammad Sobri Merais, Nurul Husna Che Hamzah, Md Bazlul Mobin Siddique, Dayangku Nurshahirah Awang Wahab, Carolina Costa, Bruno Medronho, Magnus Norgren, Håkan Edlund, Björn Lindman, Yoshitoshi Nakamura, Chikako Asada, Sholahuddin

First published in London, United Kingdom, 2021 by IntechOpen
IntechOpen is the global imprint of INTECHOPEN LIMITED, registered in England and Wales, registration number: 11086078, 5 Princes Gate Court, London, SW7 2QJ, United Kingdom
Printed in Croatia

British Library Cataloguing-in-Publication Data
A catalogue record for this book is available from the British Library

Additional hard and PDF copies can be obtained from orders@intechopen.com

Cellulose Science and Derivatives
Edited by Arpit Sand and Sangita Banga
p. cm.
Print ISBN 978-1-83969-578-0
Online ISBN 978-1-83969-579-7
eBook (PDF) ISBN 978-1-83969-580-3

For EU product safety concerns:
IN TECH d.o.o., Prolaz Marije Krucifikse Kozulić 3, 51000 Rijeka, Croatia, info@intechopen.com or visit our website at intechopen.com.

Meet the editors

Dr. Arpit Sand is currently an associate professor in the Department of Chemistry, Manav Rachna University, Faridabad, India. He received his BSc in Science and MSc in Chemistry from the University of Allahabad, India, in 2004 and 2006, respectively. He received his Ph.D. in Chemistry from the same university in 2010. Dr. Sand is an editorial board member for Polymer Synthesis Journal by Cambridge Scholars Publishing UK and *Journal of Polymer Science*. He is an academic book editor and a reviewer for international journals including *Carbohydrate Polymers*, *International Journal of Biological Macromolecules*, and *Fibers and Polymers*, among others. Dr. Sand's previous roles include assistant professor (guest faculty) in the Department of Chemistry, University of Allahabad; research associate at the National Physical Laboratory (NPL) New Delhi; postdoctoral fellow at Gwangju Institute of Science & Technology (GIST), Korea; postdoctoral fellow at Karolinska Institutet, Sweden; junior researcher at the Brno University of Technology, Czech Republic; and researcher at Soongsil University, South Korea. He has more than ten years of teaching and research experience. He is also a life member of the Indian Science Congress and Green Chemistry Network center. He has made significant contributions in the modification and characterization of graft copolymers and films decorated with chalcogenide quantum dots to tune the energy band gap for solar energy harvesting applications. His research interests include polymer synthesis using different polymerization techniques. He has authored more than twenty-four international research articles and review articles in reputed SCI journals.

Prof. Sangita obtained a master's in Chemistry with a specialization in Organic Chemistry from the Department of Chemistry, University of Roorkee (now IIT Roorkee), India. Having qualified GATE, she pursued her MTech in Polymer Science and Technology from the Indian Institute of Technology Delhi (IIT Delhi). She obtained her Ph.D. from Manav Rachna International Institute of Research and Studies (MRIIRS), India, in the area of Alternative Fuels-Biodiesel. Prof. Banga has twenty-one years of experience in teaching and research. Prof. Banga worked with an industry dealing in additives for a year in product development. She began her career in teaching as a lecturer at Lingayas Institute of Technology and Management in 2001. She joined Manav Rachna Educational Institutions in 2004 as a senior lecturer and since then has served the organization in various roles such as B. Tech Coordinator, head of the Mechanical Department, and Dean of Students & Welfare of the University. She has several publications in journals of national and international repute to her credit, and she has authored two books on engineering chemistry. Her current research work is in biofuels, mainly production from various edible, non-edible oils, algae, and purification technology using organic adsorbents, nanoparticles, and natural adsorbents. As a member of organizing committees, Dr. Banga has successfully organized the Green Chemistry workshop and Conference of the Chemistry Department in association with Green Chemistry Network Centre.

Contents

Preface

Recent developments connected to increased environmental awareness, recognition of global climate problems, and the advent of bioeconomies and biorefineries have brought cellulosic products back into the public consciousness as valuable biomaterials and chemical feedstocks. In this context, the pulp and paper industries are increasingly regarded as businesses engaged in high-tech innovation.

The development of biorefinery methods has highlighted the characteristics of celluloses such as biomineralization, recycling, conversion, and permanence. Today, the cellulose discipline is one of the most rapidly advancing fields in the natural and physical sciences. Cellulose plays and will continue to play a dominant role in the universal appearance of bioeconomies, leading the way from fossil-based chemical industries to true biorefineries. Today's uses for cellulose, tissues, fibers, and cellulose derivatives, including paper products, are only the beginning.

This book begins by addressing the conversion of biomass waste products into valuable products like cellulose hydrogel films, which are important in cell regeneration. The other chapters explain the application of cellulose derivatives in mineral processing, cellulose derivatives (CDs) that have been recognized as anionic water-soluble, non-toxic, biocompatible, and biodegradable polysaccharides. Also one chapter deals with de-inking of MOW type paper using enzymes and. One chapter covers bacterial cellulose, an alternative ecological materials have been used as viscosity regulators, thickening agents, sizing agents, coating agents, emulsion stabilizers, and electrode binders in various industries.

This book conveys the fascination of contemporary cellulose science, including its versatility, applicability, challenges, and bright future.

Arpit Sand
Department of Chemistry,
Manav Rachna University,
Faridabad, India

Sangita Banga
Manav Rachna University,
Faridabad, India

Section 1

Introduction

Chapter 1

Comparative Study of Cellulose Hydrogel Films Prepared from Various Biomass Wastes

Cho Cho, Thinzar Aye, Aung Khaing and Takaomi Kobayashi

Abstract

The conversion of biomass waste products to valuable products like cellulose hydrogel films is important in cell regeneration. In this study, the various biomass wastes: thanaka heartwood (TH), sugarcane bagasse (SB) and rice straw (RS) were used as cellulose resources. They were chemically treated using acid and alkali to obtain cellulose fibers. The yield percent of cellulose fibers depends on the nature of biomass materials. Scanning Electron Microscope (SEM), X-ray Diffraction (XRD) and Fourier Transform Infrared Spectroscopy (FTIR) analyses showed that the amount of lignin and hemicellulose from these samples were successfully reduced by chemical treatment. Cellulose fibers were treated using the dimethylacetamide/lithium chloride (DMAc/LiCl) system to obtain cellulose hydrogel solutions. Following this, the cellulose hydrogel films were prepared employing the phase inversion method without cross-linker. These films were transparent and flexible. In the present study, water retainable property and viscoelasticity of cellulose hydrogel films were measured. Antimicrobial activity tests of cellulose solutions have been carried out to be utilized to hydrogel films for biomedical application.

Keywords: cellulose hydrogel films, phase inversion method, biomass waste, antimicrobial activities

1. Introduction

Hydrogels are three-dimensional polymeric networks kept together by cross-linked covalent bonds as well as weak cohesive forces such as hydrogen or ionic bonds [1]. Hydrogels consisting of hydrophilic polymer networks can absorb up to thousands of times their dry weight in water. Natural and synthetic polymer-based hydrogels are the two types of hydrogels, classed based on polymer source. Synthetic polymer-based hydrogels are fabricated from poly ethylene glycol, poly acrylic acid, polyacrylamide, poly vinyl alcohol etc. [2–5]. Natural polymer-based hydrogels are derived from various natural polymers, such as cellulose, gelatin, peptides, chitosan and alginate, etc. [6–8] and can be employed as advanced materials for tissue and organ repair and regeneration [9–11].

Commonly, all plant biomass consists of cellulose, hemicellulose, lignin, pectin and protein with cellulose being the major component making up to 33% of plant biomass. Use of cellulose has several advantages including biocompatibility, biodegradability, renewability, good mechanical strength, being environmentally friendly, and one of the safest materials on the earth [12]. Consisting of a linear chain of β (1→4) linked *D*-glucose units, it has properties of tasteless, odorless, insolubility in water and most organic solvents - a characteristic owing to its hydrogen bonds and Van der Waals forces that make it difficult for the dissolution [13].

Abundant in hydrophilic functional groups, including hydroxyl, carboxyl, and aldehyde groups, cellulose and its derivatives are ideal materials to prepare hydrogels films. Biodegradability, biocompatibility, non-toxicity, hydrophilicity, and tissue-mimicking characteristics of cellulose-based hydrogels make them useful in a variety of sectors, including food, agriculture, environmental remediation and medicinal applications, such as drug delivery [14, 15], tissue engineering [16–19], wound dressing [20–22], bio imaging [23–25] and wearable epidermal sensors etc. [26, 27]. The possibility of these numerous biomedical applications draws researchers into exploring renewable plant biomass-based cellulose alternatives to create hydrogel films [28–31].

Myanmar is an agricultural country with paddy crop production being the main agricultural production of the country. After being harvested; rice straw is generated in the field. Sugarcane bagasse is also fibrous by-product remaining after sugar extraction from sugarcane. And, these biomass wastes are always in abundance in many areas. The bark of thanaka is used for the production of cosmetic products such as thanaka powder, liquid, and paste [32]. The heartwood is often unused and becomes waste after the bark has been used up. As a result, as sustainable cellulose, thanaka heartwood, sugarcane bagasse and rice straw are biomass wastes that can potentially be converted into a valuable product like hydrogel films and they will become cost-effective items and can reduce environmental pollution. For this purpose, this study was carried out to convert cellulose solutions from these resources by using DMAc/LiCl systems to prepare cellulose hydrogel films. This research article has also reviewed the comparative study on the properties and nature of cellulosic hydrogels from various plant biomass wastes for biomedical applications.

2. Materials and methods

2.1 Materials

Thanaka samples were obtained from Pakokku, Pakokku Township, Magway Region, Myanmar. Rice straw samples were obtained from Helgu Township, Yangon Region, Myanmar and Sugarcane bagasse collected from Nawaday Sugar Mill, Pyay Township, Bago Region, Myanmar.

2.2 Preparation of treated fiber

First, the collected raw materials were cut, washed and dried at 40°C. For the preparation of acid-treated samples, 10 g of raw samples were immersed in 500 mL, 4 vol% H_2SO_4 solution and stirred at 90°C for 2 h. After cooling down, the samples were washed, filtrated and collected as acid-treated samples. Similarly, they were again immersed in 500 mL, 10 vol % NaOH solution and stirred at 90°C. They were washed, filtrated and collected as base-treated samples. Again, they were immersed in 200 mL, 10 vol % NaOCl and stirred for the color bleaching. Then, cellulose fiber was obtained and dried at 50°C [29–31]. Percent cellulose fibers were found to be 28.50% of TH, 20.31% of SB and 21.63% of RS. Among them, the highest yield percent of cellulose was obtained from thanaka heartwood samples due to the hard portion (stem) of the plant. The yield percent of cellulose depends on the nature of plant materials.

2.3 Preparation of cellulose solution

100 mL of deionized water was added into 1 g of cellulose fiber and stirred at room temperature overnight. They are then filtered and washed with ethanol. After that, 100 mL of ethanol was added to the swollen fibers and the mixture was stirred for 24 h. Then, ethanol was removed and added to 50 mL of DMAc and stirred overnight. Finally, 8 wt % of LiCl and DMAc were added to the swollen cellulose fibers and the solvent system adjusted to obtain a 1% concentration of cellulose solution. The mixture was stirred at room temperature to get the hydrogel solution.

2.4 Preparation of cellulose hydrogel films

For hydrogel film preparation, 10 g of cellulose solution was poured into a Petri dish (10 cm diameter) and kept in a container filled with ethanol for 24 h. After this, transparent hydrogel films were obtained. Then, these films were washed with distilled water several times and immersed in distilled water and kept at room temperature. **Figures 1–3** illustrated the raw samples to cellulose hydrogel films for thanaka heartwood, sugarcane bagasse and rice straw, respectively. It was found that all cellulose hydrogel films were transparent and flexible but a little difference in their strength.

2.5 Characterization techniques

The structural changes of samples were analyzed by FTIR 8400 Shimadzu spectrophotometer by using the KBr pellet method in the MIR radiation with the wavelength from 4000 cm^{-1} to 400 cm^{-1} range with a resolution of 4.0 cm^{-1}. The surface morphology of the samples was investigated by SEM (JEOL 15 kV). The X-ray diffraction (XRD) patterns of raw sample and pretreated cellulose fibers were recorded with CuKα radiation at 40 kV and 30 mA in the range of $2\theta = 10°\text{-}40°$ by X-ray diffractometer (Smart Lab, Rigaku, Japan). The samples were dried in vacuum at room temperature before measuring them. The crystallinity index (CI) was calculated using the equation: $CI\ (\%) = (I_{002}–I_{am})/I_{002}) \times 100$, where I_{002} is the maximum intensity of the peak (002) lattice diffraction and I_{am} is the intensity of diffraction attributed to amorphous cellulose [33]. Viscoelasticity of the hydrogel film, 2 cm in diameter with 5 mm in thickness, was determined by Auto Paar-Rheoplus equipment (Anton Paar Japan, Tokyo) in wet conditions at 37°C.

2.6 Study on antimicrobial activity

The study on antimicrobial activity of DMAc, DMAc/LiCl and cellulose solutions was performed by the agar well diffusion method [34]. Nutrients agar was

(a) (b) (c) (d) (e)

Figure 1.
Photos showing (a) thanaka heartwood (b) acid-treated sample (c) base-treated sample (d) cellulose fiber and (e) thanaka heartwood cellulose hydrogel film (THCF).

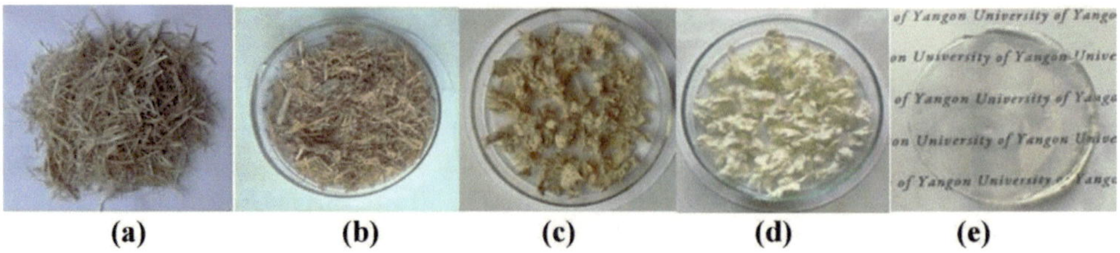

Figure 2.
Photos showing (a) sugarcane bagasse (b) acid-treated sample (c) base-treated sample (d) cellulose fiber and (e) sugarcane bagasse cellulose hydrogel film (SBCF).

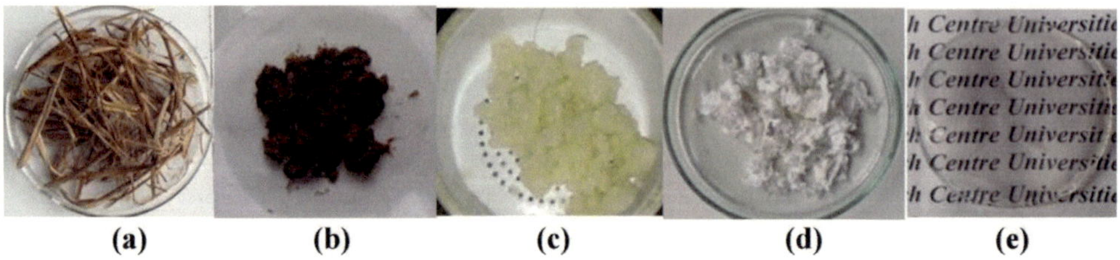

Figure 3.
Photos showing (a) rice straw (b) acid-treated sample (c) base-treated sample (d) cellulose fiber and (e) rice straw cellulose hydrogel film (RSCF).

prepared according to Cruickshank's methods [35]. Firstly, nutrient agar (medium) was boiled and 20–30 mL of the medium poured into test tubes which were plugged with cotton wool. Secondly, the test tubes with the medium were autoclaved at 121°C for 15 minutes and they were cooled down to 30–35°C. Finally, the medium was poured into the sterilized petri dishes and 0.1–0.2 mL of test organisms were added into the dishes. In this study, the tested microorganisms are *Bacillus subtilis*, *Staphylococcus aureus*, *Pseudomonas aeruginosa*, *Candida albicans* and *Escherichia coli*. The agar was allowed to set for 2–3 hours. Fourthly, 10 mm agar-well was made with the help of a sterilized agar-well cutter. Finally, about 0.2 mL of samples were introduced into the agar-well and incubated at 37°C for 24 to 36 hours. The inhibition zone that appeared around the agar-well was measured for the indication of the presence of antimicrobial activity.

3. Results and discussion

The structural changes for each biomass sample: thanaka heartwood, sugarcane bagasse and rice straw of (a) raw sample (b) acid-treated sample (c) base-treated sample and (d) cellulose fiber were shown in **Figures 4–6** respectively. The broad peaks at 3400 cm^{-1} in all samples were due to the stretching vibrations of O-H groups of water contents. The peak at 2900 cm^{-1} was due to C-H stretching. From the comparison of the presented FTIR spectra, the peaks around 1510 cm^{-1} and 1520 cm^{-1} showed the presence of lignin and lignocellulose in the initial raw sample. Moreover, the band at around 1200 cm^{-1} disappeared in treated fiber which was attributed to the removal of hemicelluloses. The changes of the bands in the region from 1423 cm^{-1} to 1080 cm^{-1} were due to the removal of lignin and hemicelluloses [36, 37]. This is a clear indication that the amount of lignin from the raw sample was successfully reduced by the chemical treatments.

The surface morphologies of biomass raw samples and their pretreated samples were shown in SEM images of thanaka heartwood in **Figure 7**, sugarcane bagasse in

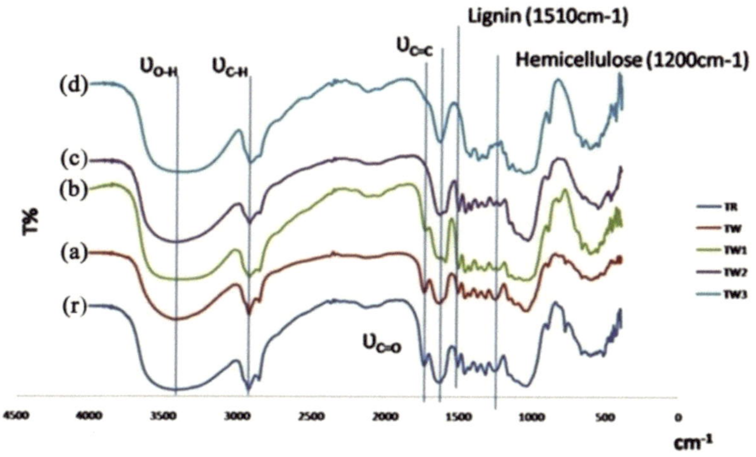

Figure 4.
FTIR spectra of (a) thanaka heartwood (b) acid-treated (c) base-treated samples and (d) cellulose fiber.

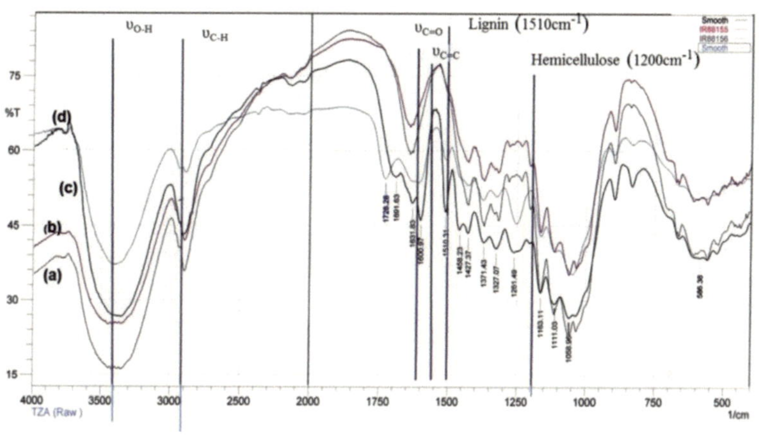

Figure 5.
FTIR spectra of (a) sugarcane bagasse (b) acid-treated (c) base-treated samples and (d) cellulose fiber.

Figure 8 and rice straw sample in **Figure 9**. Morphological changes in the raw, acid and base pretreated samples were clearly seen in these SEM images. The fibrous patterns of the plant cell wall appeared in the raw samples shown in **Figures 7(a), 8(a)**, and **9(a)**. **Figures 7(b), 8(b), 9(b)** and **7(c), 8(c), 9(c)** represented the distorted structure of lignin and hemicellulose of plant cell wall structure due to pretreatment of acid and base. Pretreatment of NaOH tended to decrease the lignin and hemicellulose which in turn increased the content of cellulose. It also resulted in an increase in cellulose accessibility shown in **Figures 7(d), 8(d)** and **9(d)**.

Water content of the hydrogel films was determined by weighing dry and wet samples by the following procedure. Disk samples with 5 mm diameter were cut from cast films and dried in a vacuum oven for 24 h and weighed (W_0). Then, samples were immersed in distilled water for 36 h. After that, films were removed from the water and wrapped with filtered paper in order to remove excess water and weighed again (W_1). Finally, the equilibrium water content was calculated from the wet (W_1) and dried (W_0) hydrogel films. For each sample, four independent measurements were done and averaged by the following formula: Water content (%) = $(W_1–W_0)/W_0 \times 100$. According to the calculation, water content was 165.5%

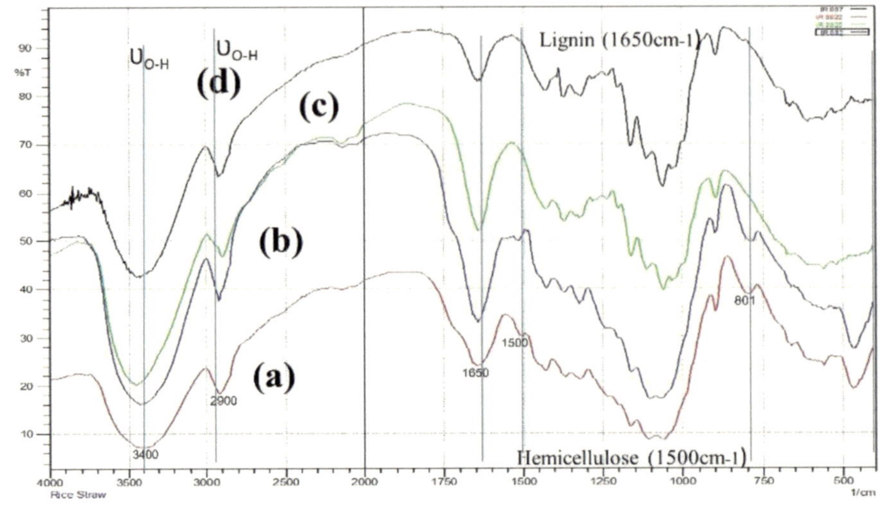

Figure 6.
FTIR spectra of (a) rice straw raw (b) acid-treated samples (c) base-treated and (d) cellulose fiber.

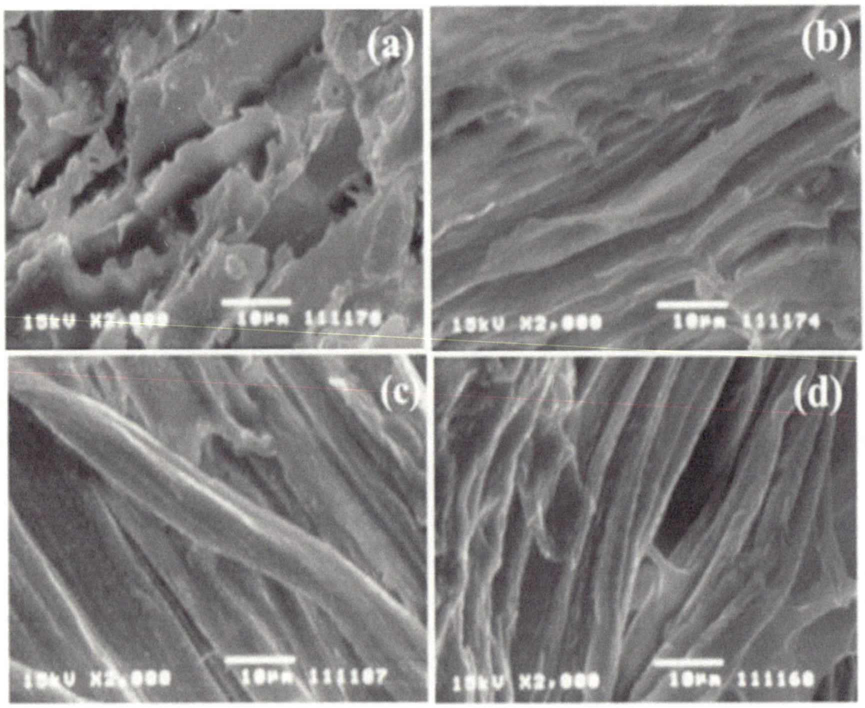

Figure 7.
Comparative SEM images of (a) thanaka heartwood (b) acid-treated (c) base-treated samples and (d) cellulose fiber.

for THCF, 188.47% for SBCF film and 168.63% for RSCF film. Thus, the films have water retainable property which will be the formation of the interaction of hydrogen bonding networks of the resultant cellulose in the hydrogel films.

XRD measurement was carried out to evaluate the effect of treatment conditions on the crystalline structure of raw samples, treated fibers and cellulose fibers. The patterns of (a) to (d) exhibited typical crystalline lattice of cellulose with peaks at 22.3° and 16.4° [38, 39]. This cellulose crystalline can be found in natural

Figure 8.
Comparative SEM images of (a) sugarcane bagasse (b) acid-treated (c) base-treated samples and (d) cellulose fiber.

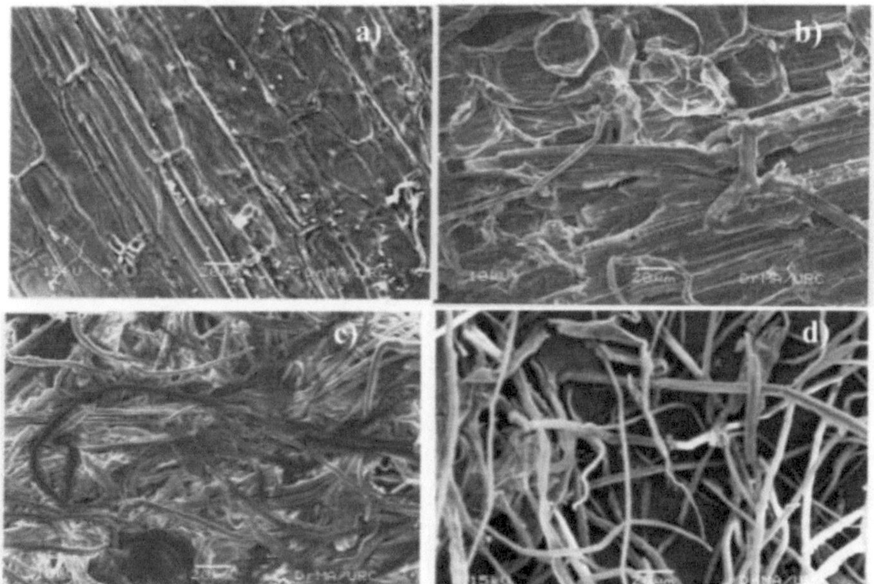

Figure 9.
Comparative SEM images of (a) rice straw (b) acid-treated (c) base-treated samples and (d) cellulose fiber.

plant cellulose. **Figure 10** showed the XRD patterns of thanaka heartwood and the treated fibers. In the crystallinity of thanaka heartwood and cellulose fibers, the peak ratio showed that thanaka sample and pre-treated fibers were at 68.5%, 82.1%, 84.2%, and 85.4%, respectively. **Figure 11** showed the XRD patterns of the bagasse

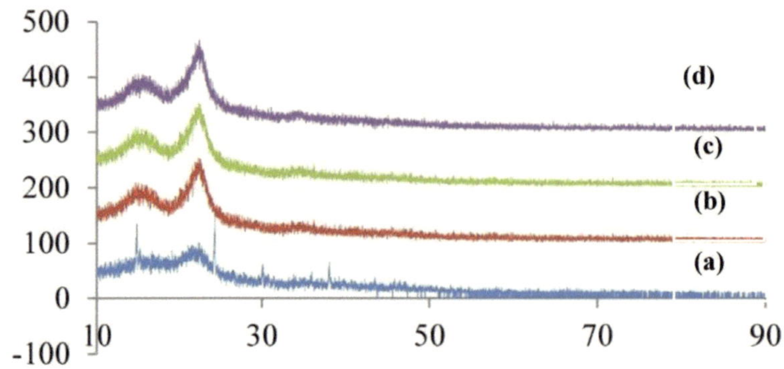

Figure 10.
XRD patterns of (a) thanaka heartwood (b) acid-treated (c) base-treated and (d) cellulose fiber.

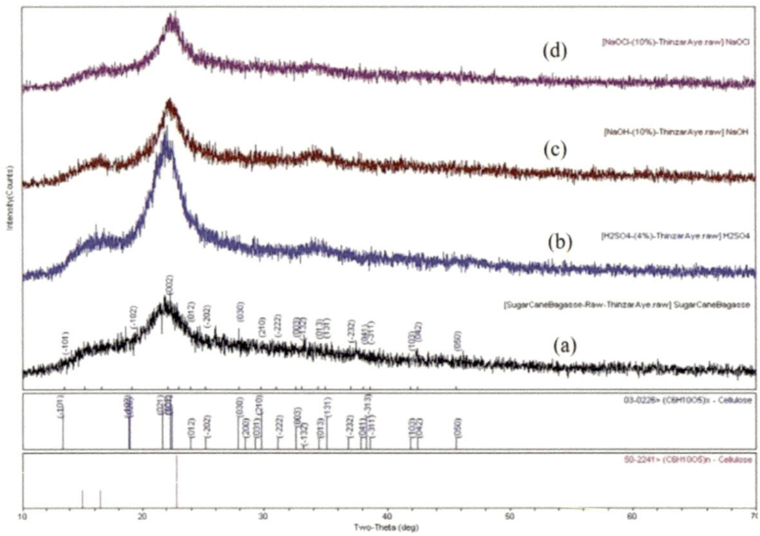

Figure 11.
XRD patterns of (a) sugar cane bagasse (b) acid-treated (c) base-treated and (d) cellulose fiber.

and the purified fibers. The crystallinity indexes of sugar cane bagasse raw sample, acid-treated sample, base-treated sample and cellulose fiber were 44.1%, 58.8%, 59.1% and 60.2% respectively. **Figure 12** showed the XRD patterns of rice straw and the purified fibers. The crystallinity indexes of rice straw, acid-treated sample, base-treated sample and cellulose fiber were 71.43%, 75.7%, 76.25% and 78.57%, respectively. The increment of the crystallinity in the pre-treated fibers was due to the removal of hemicelluloses and lignin by sodium hydroxide and sodium hypochloride treatment, indicating especially higher purity of cellulose fibers.

Viscoelasticity measurements indicated the relationship between storage elastic modulus G', loss elastic modulus G" and strain for the hydrogel films. **Figure 13** showed the viscoelastic data for (a) THCF films (b) SBCF films and (c) RSCF films. It was noted that the deformation of THCF occurred at G'= 8×10^4 and G"= 8×10^3 Pa at 1×10^{-2} to 9% strain, SBCF at G'= 7×10^4 and G"= 6.5×10^3 Pa at 1×10^{-2} to 35.71% strain, the RSCF film at G'= 8×10^4 and G"= 7.5×10^3 Pa at 1×10^{-2} to 12.45% strain [40]. The loss elastic modulus (G") of SBCF film was lower than that of the others. This meant that the elastic nature was low and deformation was high in SBCF film. The crossover point of G' and G" meant fracture of materials or inability to follow deformation due to the rigid polymer network. In the prepared

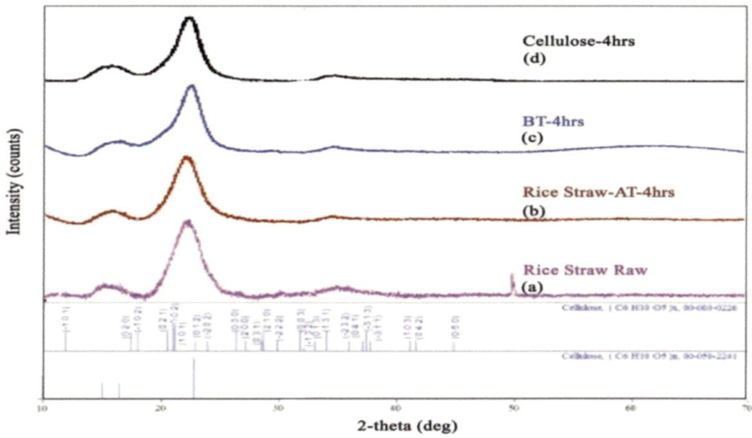

Figure 12.
XRD patterns of (a) rice straw raw (b) acid-treated (c) base-treated and (d) cellulose fiber.

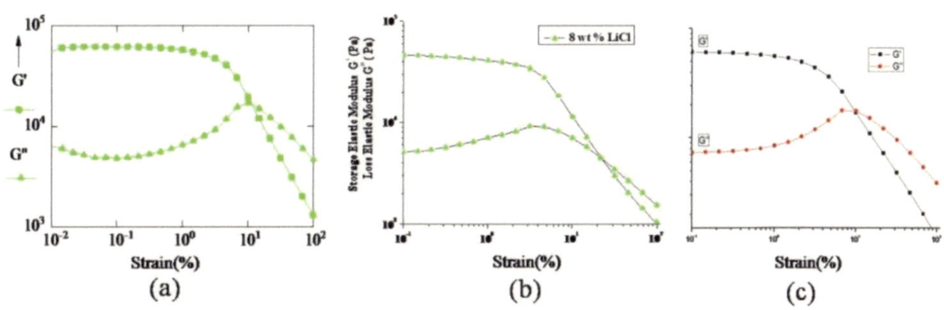

Figure 13.
Viscoelasticity of (a) thanaka heartwood cellulose hydrogel film (THCF), (b) sugarcane bagasse cellulose hydrogels films (SBCF) and (c) rice straw cellulose hydrogel film (RSCF).

cellulose hydrogel, cross points of G' and G" shifted towards a low strain range in order of strength of cellulose fiber. The crossover point of THCF film was found to be 9% and that of RSCF was 12.45% strain. In the case of SBCF, the G' and G" values overlapped at 35.71% strain. It seemed that the SBCF film showed a very soft and less elastic nature than the other. Based on the comparison of viscoelasticity measurement for these three films, THCF film was the strongest and most elastic compared to the other two films as a consequence of its rigid polymer network of hardwood resource.

Table 1 showed the measurements of inhibition zones of cellulose solutions. The inhibition zone appearing around the agar-well was measured; an indication of the presence of antimicrobial activity. The measurable zone diameter, including the well diameter showed the degree of antimicrobial activity. The well diameter was 10 mm in this experiment. The larger the inhibition zone diameter showed stronger anti-microbial activity on the test organisms. According to the results, cellulose solution samples showed an inhibition zone more than the DMAc/LiCl solution, noting the DMAc solvent had no antimicrobial activity. It was found that all cellulose solutions showed antimicrobial activity and TH cellulose solution exhibited high activity against all test microorganisms (diameter of inhibition zone ranging 12–18 mm). On grounds of the antimicrobial activity results and water retaining properties, cellulose hydrogel films prepared from plant biomass waste can be applied as wound dressing, medical plaster and facial mask for biomedical applications.

Samples	*Bacillus subtilis*	*Staphylococcus aureus*	*Pseudomonas aeruginosa*	*Bacillus pumilus*	*Candida albicans*	*Escherichia coli*
DMAc	—	—	—	—	—	—
DMAc + LiCl	15 mm (+)	15 mm (+)	15 mm (+)	15 mm (+)	12 mm (+)	15 mm (+)
RS cellulose solution	17 mm (++)	15 mm (+)	16 mm (++)	17 mm (++)	15 mm (+)	15 mm (+)
SB cellulose solution	16 mm (++)	15 mm (+)	16 mm (++)	17 mm (++)	13 mm (+)	15 mm (+)
TH cellulose solution	18 mm (++)	18 mm (++)	15 mm (+)	17 mm (++)	12 mm (+)	18 mm (++)

Table 1.
Antimicrobial activity results for test samples.

4. Conclusion

By using DMAc/LiCl system, conversion of biomass waste to valuable products like cellulose hydrogel films was successfully prepared. Comparative studies on the properties and nature of cellulosic hydrogels from various plant biomass wastes were also reviewed. The yield percent of cellulose fibers: 28.50% of thanaka heartwood, 20.31% of sugarcane bagasse and 21.63% of rice straw was found in the study. It was regarded that the amount of the cellulose fibers depends on the nature of biomass materials. Cellulose hydrogel films were transparent and flexible but of slightly variable strengths due to the different kinds of cellulose sources. The water retainable property of SBCF was greater than that of THCF and RSCF. Based on the comparison of viscoelasticity measurement, it was found that THCF and RSCF were stronger, more elastic in nature and of a less deformed nature than the SBCF film. Since the cellulose solutions displayed antimicrobial activities against six test organisms, the cellulose hydrogel films could be utilized in biomedical applications.

Acknowledgements

We are extremely thankful to experts at Open Access Funding Page and especially to Mia Vulovic, Author Service Manager, for granting a free publishing fee allowing us to share our research with a wider audience. The authors: Cho Cho, Thinzar Aye and Aung Khaing, would like to express our sincere thanks to Prof. Dr. Pho Kaung, Rector of the University of Yangon for motivating and heartening us in our research. We are greatly appreciated Prof. Dr. Ye Chan, Head of Universities' Research Centre (URC), for his permission to carry out this research. We are deeply indebted to Prof. Dr. Takaomi Kobayashi, Department of Science of Technology Innovation, Nagaoka University of Technology, Nagaoka, Japan, for his invaluable advice and endless support. Our special thanks go to all of the URC staff.

Conflict of interest

The authors declare no conflict of interest.

Author details

Cho Cho[1]*, Thinzar Aye[2], Aung Khaing[1] and Takaomi Kobayashi[3]

1 Department of Chemistry, University of Yangon, Yangon, Myanmar

2 Department of Chemistry, Pyay University, Bago Region, Myanmar

3 Department of Science of Technology Innovation, Nagaoka University of Technology, Nagaoka, Japan

*Address all correspondence to: chocho@uy.edu.mm

References

[1] Stanisław M, Alina S, Amit J: Biopolymers for hydrogels in cosmetics. Journal of Materials Science: Materials in Medicine. 2020;31:50 doi:10.1007/s10856-020-06390-w

[2] Jeong B, Kim SW, Bae YH: Adv. Drug Delivery Rev., 2012;64:154-162.

[3] Ahmed EM. J: Adv. Res. 2015;6: 105-121

[4] Xu B, Zheng P, Gao F, Wang W, Zhang H, Zhang X, Feng X, Liu W: Adv. Funct. Mater. 2017;27:1604327

[5] Ge G, Zhang Y, Shao J, Wang W, Si W, Huang W, Dong X: Adv. Funct. Mater. 2018;1802576, DOI:10.1002/adfm.201802576

[6] Alvarez-Lorenzo C, Blanco-Fernandez B, Puga AM, Concheiro A: Adv. Drug Delivery Rev., 2013;65: 1148-1171.

[7] De France KJ. Hoare T. Cranston ED: Chem. Mater. 2017;29:4609-4631.

[8] Dou XQ, Feng CL: Adv. Mater. 2017;29:1604062

[9] Ammini K, Karthika S, Sindhu A, Vijay RP, Thakur K, Thakur VK: Medical Applications of Cellulose and its Derivatives: Present and Future. In Book: Nanocellulose Polymer Nanocomposites: Fundamentals and Applications, Wiley; 2014.437-473. DOI:10.1002/9781118872246.ch16

[10] Kamide K: Cellulose and cellulose derivatives molecular characterization and its applications. 1st ed. Elsevier Science 2005

[11] Kobayashi T: Fabrication of cellulose hydrogels and characterization of their biocompatible films. In studies in natural products chemistry. Amsterdam, Netherlands: Elsevier B.V. 2017;45:1-15

[12] Chen X, Chen J, You T, Wang K, Xu F: Effects of polymorphs on dissolution of cellulose in NaOH/urea aqueous solution. Carbohydrate Polymers. 2015;a,b;125;85-91

[13] Luo X, Zhang L: New solvents and functional materials prepared from cellulose solutions in alkali/urea aqueous system. Food Resources International. 2010;52(1);387-400.

[14] Hujaya SD, Lorite GS, Vainio SJ, Liimatainen H: Acta Biomater., 2018;75:346-357

[15] Pakulska MM, Vulic K, Tam RY, Shoichet MS: Adv. Mater. 2015;27: 5002-5008

[16] Svensson A, Nicklasson E, Harrah T, Panilaitis B, Kaplan DL, Brittberg M, Gatenholom P: Bacterial cellulose as a potential scaffold for tissue engineering of cartilage, Biomaterials. 2005;26:419-425.

[17] Karla LT, Satoshi SS, Tagaya M, Kobayashi T: Fibroblast compatibility on scaffold hydrogels prepared from agave tequilana weber bagasse for tissue regeneration. Ind. Eng. Chem. Res. 2013;52:11607-11613.

[18] Wang K, Nune KC, Misra RD: Acta Biomater. 2016;36:143-151

[19] Boyer C, Figueiredo L, Pace R, Lesoeur J, Rouillon T, Visage CL, Tassin JF, Weiss P, Guicheux J, Rethore G: Acta Biomater. 2018;65:112-122

[20] Pinho E, Soares G: J. Mater. Chem. B, 2018;6:1887-1898

[21] Huber D, Tegl G, Mensah A, Beer B, Baumann M, Borth N, Sygmund C, Ludwig R, Guebitz G M: ACS Appl. Mater. Interfaces, 2017;9:15307-15316

[22] Mao C, Xiang Y, Liu X, Cui Z, Yang X, Yeung KWK, Pan H, Wang X,

Chu PK, Wu S: ACS Nano. 2017;11: 9010-9021

[23] Chang C, Peng J, Zhang L, Pang DW: J. Mater. Chem. 2009;19:7771

[24] Wang Z, Fan X, He M, Chen Z, Wang Y, Ye Q, Zhang H, Zhang L: J. Mater. Chem. B. 2014;2:7559-7566

[25] Hai J, Li T, Su J, Liu W, Ju Y, Wang B, Hou Y: Angew. Chem., Int. Ed. 2018;57:6786-6790

[26] Liu YJ, Cao WT, Ma MG, Wan P: ACS Appl. Mater. Interfaces. 2017;9:25559-25570

[27] Shao C, Wang M, Meng L, Chang H, Wang B, Xu F, Yang J, Wan P: Chem. Mater., 2018;30:3110-3121

[28] Kazuki N, Kobayashi T: Effect of pre-treatment of sugar cane bagasse on the cellulose solution and application for the cellulose hydrogel films. J. Polym. Adv. Technol. 2016;27:973-980

[29] Cho Cho, Kobayashi T. Preparation and Characterization of Cellulose Hydrogel Films from Myanmar Thanaka Heartwood. Transaction on GIGAKU. 2018;5(1)05004/1-8

[30] Cho Cho, Cho Cho Thet, Nan Sanda Myint, Wut Yee Aye: Cellulose hydrogel films prepared from biomass waste Rice straw. Universities' research Journal. 2017;10:2:55-66

[31] Thinzar Aye, Thinzar Nu, Cho Cho: Preparation of cellulose hydrogel films from sugarcane bagasse. J. Myanmar Acad. Arts Sci. 2018;XVI; No. 1B:371-378

[32] https://yangonlife.com.mm/en/article/thanaka-secret-burmese-beauty[Internet]. 2015.

[33] Azubuike CP, Rodrígue H, Okhamafe AO: Physicochemical, spectroscopic and thermal properties of microcrystalline cellulose derived from corn cobs. International journal of recycling of Organic Waste in Agriculture. 2012;19:425-433

[34] Savoia D. Plant-derived antimicrobial compounds: Alternatives to antibiotics. Future Microbial. 2012;7(8):979-990.

[35] Cruickshank R, Duguid JP, Marmion BP, Swain RHA: Medical Microbiology. NewYork. Churchill Livingstone. 1975:2.

[36] Langkilde FW, Skantesson A: Identification of celluloses with Fourier-transform (FT) mid-infrared, FT-Raman and near-infrared spectroscopy. Journal of Pharmaceutical and Biomedical Analysis. 1995;13:409-415,1995.

[37] D. Fengel: Characterization of cellulose by Deconvoluting the OH Valency range in FTIR spectra. Holzforschung Journal. 1992;46:No. 4:283-288

[38] Nishino T, Matsuda Y, Hirao K: Macromolecules, 2004;37:7683-7687.

[39] Lindman B, Medronho B, Alves L, Costa C, Edlund H, Norgren M: The relevance of structural features of cellulose and its interactions to dissolution, regeneration, gelation and plasticization phenomena. Physical Chemistry Chemical Physics. 2017; 19: 23704-23718. DOI:10.1039/C7CP02409F.

[40] García MC, Alfaro MC, Calero N, Muñoz J: Influence of gellangum concentration on the dynamic viscoelasticity and transient flow of fluid gels. Biochemical Engineering Journal. 2011;55:73

Chapter 2

Application of Cellulose Derivatives in Mineral Processing

Ashok Kumar, Kaman Singh and Satya Prakash Gupta

Abstract

Cellulose derivatives (CDs) have been recognized as an anionic water-soluble, non-toxic, biocompatible and biodegradable polysaccharide. The CDs have been used as a viscosity regulator, thickening agent, sizing agent and coating agent and emulsion stabilizer, electrode binder in various industries. These characteristics properties of CDs are associated with hydroxyl groups/functionalized groups present in their structure. The CDs have significant advantages in various fields including several industrial applications such as mineral processing, palletisation process, oil drilling industrial applications due to their non-toxic and selective properties. Moreover, The CDs have been extensively used as a depressant, dispersant as well as flocculants in mineral processing from various ores. During the mineral processing like flotation of sulfide minerals highly toxic inorganic species were used as dispersant and depressant which ultimately cases environmental toxicity. Therefore, there is a current need to introduce CDs as various alternative nontoxic dispersant and flocculants. This chapter emphasized an overview of the application of CDs in mineral processing including the structure, properties of the commonly used minerals processing.

Keywords: cellulose derivatives (CDs), mineral processing, depressant, dispersant

1. Introduction

Cellulose has been recognized as the most abundant polymer on the planet, making it an important raw material for a variety of applications. Because of its potential use in the development of biofuels, cellulose has recently gained attention. However, cellulose's flexibility has been demonstrated in a variety of applications. It can also be chemically modified to produce cellulose derivatives (CDs) [1]. The two main classes of cellulose derivatives (CDs) were Cellulose ethers and cellulose esters, which have different physicochemical and mechanical properties. The Cellulose derivatives (CDs) have been widely used in a wide range of applications, including particle dispersion, flocculation processes, surface treatment, and so on. Tablet binding, thickening, film-forming, water retention, adhesion, and suspending and emulsifying agents are some of the most common uses of cellulose derivatives in tablet and capsule formulations. Natural aggregates are still important in cement production, though high-purity sources are becoming harder to come by. The CDs have been used to "inert" the aggregate to the cement formulation to prevent clay minerals associated with the aggregates from adversely affecting the cement formulation by adsorption of plasticizers and resulting property alteration [2]. The CD's application in the upstream petroleum industry, such as exploration, drilling, development, and distribution, has recently sparked renewed interest. Adding CDs to fluids, in particular, can have important benefits for improved oil recovery and well drilling, such as changing fluid properties, rock wettability alternation, advanced drag reduction, sand consolidation strengthening, minimizing interfacial friction, and increasing mobility of capillary-trapped oil [3].

For the depression of copper minerals, inorganic modifiers such as sodium cyanide, sodium sulphide or hydrosulphide, ferrocyanides, and Nokes reagent are frequently used. These reagents were very reliable, but their use has recently raised environmental concerns. CDs depressants have been investigated as possible alternatives to avoid this issue. The use of polysaccharides namely starch, dextrin, and carboxymethylcellulose (CMC) in sulphide minerals have been identified [4–11]. A frother is produced by combining hydroxypropylmethylcellulose (HPMC) or hydroxyethyl methylcellulose (HMC) with at least one non-ionic organic surfactant or polyglycol esters. The new cellulose-based frothers can be used in mineral processing plants to allow for the processing of larger amounts of minerals without requiring major changes to existing equipment. This book chapter describes the structure, properties of the commonly used CDs in minerals processing.

2. Background

Raw material demand has steadily increased on a global scale as a result of demographic and economic developments. If current trends in raw material use intensify, Industrial technologies will be unable to meet this growing demand. As a result, it is necessary to highlight that raw material (CDs present case) production must be supported as a strategic necessity, requiring the development of new technologies that can help meet this upcoming raw material demand. An additional challenge for the mining industry is that, as the world's mineral reserves are exhausted and demand for metallic raw materials rises, it will be designed to process ever greater amounts of low-quality mined material to manufacture concentrates in adequate quantities to meet current and potential demand. As a result of these requirements, minerals trapped in tailings ponds have started to receive interest as a potential source of raw materials, as many existing ore bodies near depletion. As a result, the mineral processing industry was interested in finding new solutions for the treatment of tailings to reprocess. It's worth remembering that the global amount of tailings was enormous, and if an available processing method could be developed, this could translate into a massive feed stream for the metals industry [11]. One of the most commonly used enrichment methods in mineral processing is froth flotation separation. The flotation separation is used in the processing of metals like copper, gold, and platinum to produce concentrates that can be refined economically. Efforts are currently being made to better understand the various phenomenon like frother phenomenon, adsorption etc. and how to regulate it, but no genuinely innovative frother method has been proposed, as most studies are focusing on CDs [12, 13]. As soon as synthetic polymers (CDs) were introduced into mineral processing, combinations with cellulosic materials became available and the importance of CDs came to light.

Selective depressants were essential components of any flotation reagent scheme that aims to separate various minerals selectively. Inorganic depressants have been used often. Many of these depressants, especially those used in differential sulphide flotation, are highly toxic and unsuitable for use in the environment. Sodium cyanide, sodium dichromate, sulfur dioxide, arsenic trioxide, phosphorous pentasulfide, and other depressants were examples. Since some of these inorganic depressants were reducing agents, they can be oxidized in aerated flotation pulps. The use of a lot of reagents was normally the result. Polysaccharides, on the other hand, were non-toxic and biodegradable natural organic polymers. They're much less expensive and less prone to oxidation than inorganic depressants. These characteristics not only make them perfect flotation reagents, but they have also shown promise as selective depressants in a variety of differential mineral flotation systems. For nearly 70 years, cellulose derivatives (CDs) and polysaccharides/CDs have been used in the mineral industry as depressants for iron oxides, naturally hydrophobic minerals, and rock-forming gangue minerals. They've also been reported to be selective in sulphide mineral differential flotation [14].

3. Cellulose derivatives (CDs)

Cellulose is a linear polymer made up of D-glucose monomers linked together by D-β (1–4) linkages and arranged in repeating cellobiose units, each of which contains two anhydroglucoses (**Figure 1**). Cellulose has a long molecular chain and the three hydroxyl groups have a high hydrogen-bonding ability. The hydrogen atoms of hydroxyl groups in cellulose's anhydroglucose units were replaced with alkyl or substituted alkyl groups to create cellulose ethers, which have a high molecular weight. The molecular weights, chemical structure, and distribution of substituent groups, as well as the degree of substitution and molar substitution, determine the commercially important properties of cellulose ethers (where applicable) [1]. The solubility, viscosity in solution, surface activity, thermoplastic film characteristics, and resilience against biodegradation, heat, hydrolysis, and oxidation were all examples of these properties. The molecular weights of cellulose ether solutions were specifically correlated to their viscosity. Methylcellulose (MC), ethyl cellulose (EC), hydroxyethyl cellulose (HEC), hydroxypropyl cellulose (HPC), hydroxypropylmethylcellulose (HPMC), carboxymethyl cellulose (CMC), and sodium carboxymethyl cellulose have several identified of the most commonly used sodium-carboxyl methylcellulose (Na-CMC). However, the CDs HPMC, HPC, microcrystalline cellulose (MCC), silicicedmicrocrystallinecellulose (SMCC), HEC, sodium carboxymethylcellulose (SCMC), ethylcellulose (EC) methylcellulose (MC), oxycellulose (OC), etc. have also been used in allied industries **Table 1** [14, 15].

3.1 Carboxymethylcellulose (CMC)

Carboxymethyl cellulose (CMC) has been introduced as a cellulose derivative in which some of the hydroxyl attached to them (-CH2-COOH) make up the cellulose backbone (**Figure 2**). The alkali-catalyzed reaction of cellulose with chloroacetic acid produces it. The polar carboxyl groups in cellulose (organic acid) rendering it soluble and chemically reactive. The degree of substitution of the cellulose structure (i.e., how many of the hydroxyl groups have taken part in the substitution

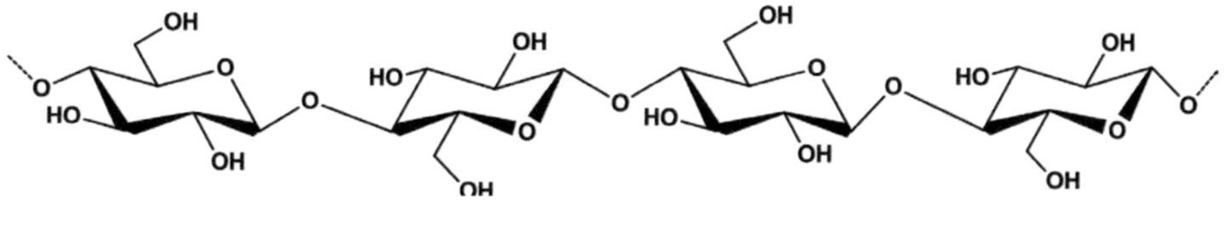

Figure 1.
Structure of cellulose (repeating unit of glucose) shows the d-glucose units are linked through β-1, 4 bonds. ----O represents the continuation of the polymeric chain.

Cellulose ethers		
Category	**Group R = H or**	**Derivatives**
Alkyl	$-CH_3$	Methylcellulose (MC)
	$-CH_2CH_3$	Ethylcellulose (EC)
	$-CH_3$ or $-CH_2CH_3$	Ethylmethylcellulose (HMC)
Hydroxyalkyl	$-CH_2CH_2OH$	Hydroxyethylcellulose (HEC)
	$-CH_2CH(OH)CH_3$	Hydroxypropylcellulose (HPC)
	$-CH_3$ or $-CH_2CH_2OH$	Hydroxyethylmethylcellulose (HEMC)
	$-CH_3$ or $-CH_2CH(OH)CH_3$	Hydroxypropylmethylcellulose (HPMC)
	$-CH_2CH_3$ or $-CH_2CH_2OH$	Ethylhydroxyethylcellulose (EHC)
Carboxyalkyl	$-CH_2COOH$	Carboxymethylcellulose (CMC)
Cellulose ester		
Organic	H or $-(C{=}O)CH_3$	Celluloseacetate
	$-(C{=}O)CH_3$	Cellulosetriacetate
	H or $-(C{=}O)CH_2CH_3$	Cellulosepropionate
	H or $-(C{=}O)CH_3$ or $-(C{=}O)CH_2CH_3$	Celluloseacetatepropionate (CAP)
	H or $-(C{=}O)CH_3$ or $-(C{=}O)CH_2CH_2CH_3$	Celluloseacetatebutyrate (CAB)
Inorganic	H or $-NO_2$	Nitrocellulose (cellulose nitrate)
	H or $-SO_3H$	Cellulose sulfate

Table 1.
Some most common class of cellulose derivatives (CDs).

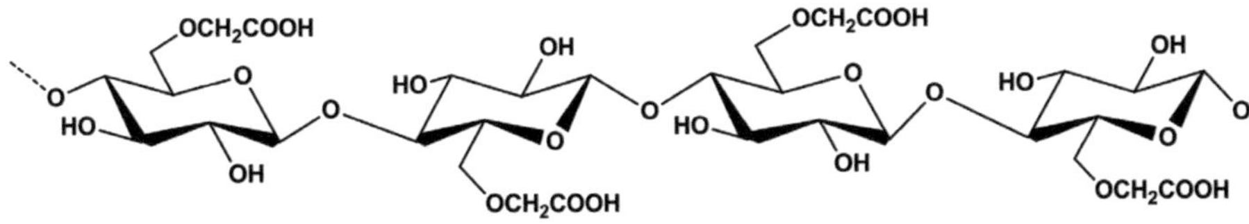

Figure 2.
Structure of Carboxymethulcellulose (repeating unit of glucose) shows the d-glucose units are linked through β-1, 4 bonds. ----O represents the continuation of the polymeric chain.

reaction), as well as the chain length of the cellulose backbone structure and the degree of clustering of the Carboxymethyl substituents, impact the functional properties of CMC.

The CMC was also used in the oil drilling industry as a viscosity modifier and water-retaining agent in drilling mud. CMC has been used to make poly-anionic cellulose (PAC) which was often used in oilfield operation. Some researchers performed surface modification and used surfactant to adjust the surface tension of the carbon fiber to improve dispersion. The wettability of carbon fibers by water was effectively improved by ozone surface treatment, which increases the fiber-matrix bond [15]. The silane treatment of carbon fibers enhances the mechanical properties of carbon fiber reinforced cement, according to Xu and Chung [16]. Wang et al. [17] used hydroxyethyl cellulose and an ultrasonic wave to support fiber dispersion in carbon fiber-reinforced cement-based composites. As a dispersing agent, CMC was used. CMC can increase carbon fiber dispersion because it has both hydrophobic and hydrophilic sides as a dispersant. Concerning carbon fiber dispersion, the effects of CMC concentration and solution pH were investigated [18].

3.2 Sodiumcarboxymethylcellulose (CMC)

Sodium carboxymethyl cellulose (CMC) is one of the most important products of cellulose ethers, which have been cellulose derivate with an ether structure produced by natural cellulose modification (**Figure 3**). Since the acid form of CMC has a low water solubility, it is generally preserved as sodium carboxymethylcellulose, which is widely used in many industries and is commonly referred to as monosodium glutamate [19, 20].

Na-CMC can be supplied stably and in large quantities, and its technical and cost efficiency was uniform and robust as compared to botanical natural polysaccharides such as tragacanth, arabic, and gua gums, or microbiological polysaccharides such as xanthan gum, which perform the same functions. Na-CMC-Na can thus be recommended as a suitable additive for coal-water slurries (CWS) as an energy supply source with low cost, stability, uniform property, and broad quantity supply capability [21]. A bituminous coal sample from Zonguldak (Thermal Code No. 434) was also used in addition to the brown coals. In all of the samples, Na-CMC was used as a stabilizer in the preparation of CWS [22]. Besides this Na-CMC was used to relieve dry, irritated eyes. Common causes for dry eyes include wind, sun, heating/air conditioning, computer use/reading, and certain medications. Electronics, pesticides, leather, plastics, printing, ceramics, and the daily-use chemical industry are only a few of the fields where Na-CMC has been used.

3.3 Hydroxyethyl cellulose (HEC)

Hydroxyethylcellulose (HEC) (**Figure 4**) has been introduced a polysaccharide derivative with gel thickening, emulsifying, bubble-forming, water-retaining and stabilizing properties. It is a white, yellowish-white or grayish-white, odorless and tasteless, hygroscopic powder [23, 24].

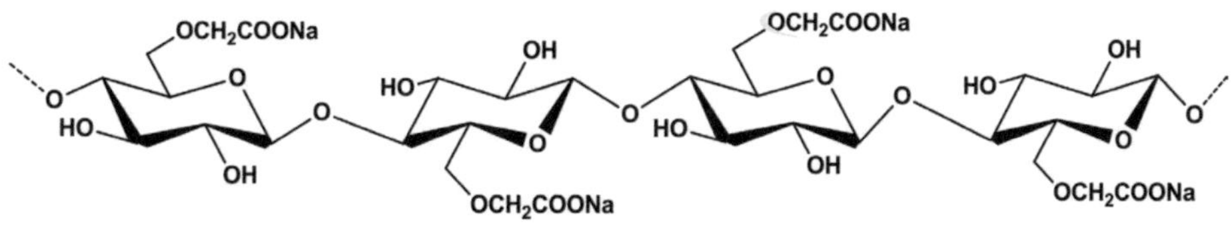

Figure 3.
Structure of sodium carboxymethylcellulose (repeating unit of glucose) shows the d-glucose units are linked through β-1, 4 bonds. ----O represents the continuation of the polymeric chain.

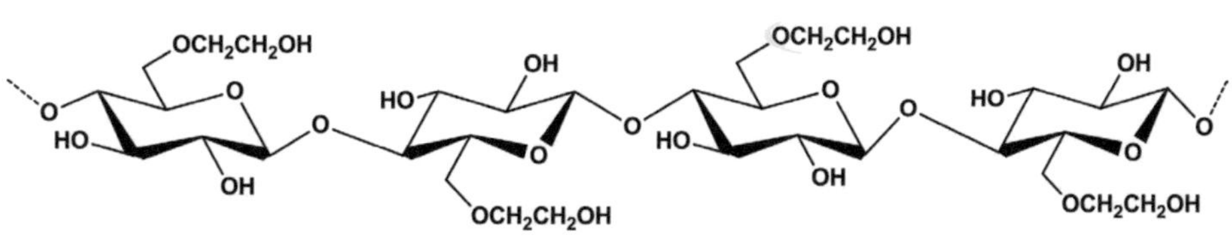

Figure 4.
Structure of hydroxyethylcellulose (repeating unit of glucose) shows the d-glucose units are linked through β-1, 4 bonds. ----O represents the continuation of the polymeric chain.

Hydroxyethylcellulose (HEC) and its hydrophobically modified derivatives have been widely used in many industrial areas such as pharmaceuticals, cosmetics, textiles, paint and mineral industries including in oil extraction, coating, medication, food, and polymer polymerization. This was non-toxic and inexpensive. Despite this, only a few reports on HEC as a sulphide depressant have been published. The function of hydroxyethyl cellulose (HEC) in the flotation separation of chalcopyrite and galena has been investigated, and the explanation for this has been explained. In the presence of H2O2, a small amount of HEC can depress galena flotation but had only a negligible effect on chalcopyrite flotation. HEC was adsorbed on galena surfaces primarily through chemical reactions with oxidation products formed on the surface, and the addition of H2O2 can significantly improve HEC adsorption by producing further oxidation products. As a result, its research would be used as a galena depressant in the flotation separation of chalcopyrite and galena, as well as to propose a method for separating copper/lead sulphide minerals. Furthermore, HEC can be used as the stabilizer of beer foam [25–30].

4. Conclusions

Mining is one of humanity's oldest industries, and it has aided in the advancement of technology that has brought us our modern products. The rising demand for mineral-based technological goods, combined with the increasingly difficult and complex method of obtaining raw materials, necessitates more efficient separation and recycling processes. At the same time, the pressures on mineral processing in terms of sustainability and environmental friendliness are making it more difficult to produce minerals as a scarce resource economically viable. Therefore, Cellulose derivatives (CDs), high-molecular-weight condensation polymers made up of basic monosaccharide sugar units, has been applied in mineral processing. Although there were many different types of polysaccharides including their derivatives in nature, wildly have been used in the mineral industry, especially in flotation. The chapter which is presented here is a small contribution to the growing understanding of polymer/CD's applications. We thus feel that this chapter is a valuable contribution to the field of mineral processing and its allied areas with a brief description of CDs. Hence in this chapter, we have summarized the current status of important cellulose derivatives employed in mineral processing.

Author details

Ashok Kumar[1], Kaman Singh[1*] and Satya Prakash Gupta[2]

1 Surface Science Laboratory, Department of Chemistry, Babasaheb Bhimrao Ambedkar University (A Central University), Lucknow, U.P., India

2 Prasad Polytechnic Lucknow, Sarai Sahjadi, Uttar Pradesh, India

*Address all correspondence to: singh.kaman@bbau.ac.in; drkamansingh@yahoo.com

References

[1] Shokri, J., & Adibki, K. (2013). Application of Cellulose and Cellulose Derivatives in Pharmaceutical Industries. Cellulose - Medical, Pharmaceutical and Electronic Applications. DOI:10.5772/55178

[2] Jacquet, A., Geatches, D., Clark, S., & Greenwell, H. (2018). Understanding Cationic Polymer Adsorption on Mineral Surfaces: Kaolinite in Cement Aggregates. Minerals, 8(4), 130. Doi: 10.3390/min8040130

[3] Cheraghian, G., Khalili Nezhad, S. S., Kamari, M., Hemmati, M., Masihi, M., & Bazgir, S. (2014). Adsorption polymer on reservoir rock and role of the nanoparticles, clay and SiO2. International Nano Letters, 4(3). DOI: 10.1007/s40089-014-0114-7.

[4] Rath, R. K., Subramanian, S., Sivanandam, V., & Pradeep, T. (2001). Studies on the Interaction of Guar Gum With Chalcopyrite. Canadian Metallurgical Quarterly, 40(1), 1-11. DOI:10.1179/cmq.2001.40.1.1.

[5] D.R. Nagaraj, S.S. Wang, P.V. Avotins and E. Dowling,. Trans. IMM-Minerals Processing and Extractive Metallurgy: Section C. Vol. 95, p. 17 (1986)

[6] J.R. Schnarr, Milling Practice in Canada. CIM Spec. Vol. 16, p. 158 (1978).

[7] Laskowski, J., Liu, Q., & Bolin, N. (1991). Polysaccharides in flotation of sulphides. Part I. Adsorption of polysaccharides onto mineral surfaces. International Journal of Mineral Processing, 33(1-4), 223-234.

[8] K.G. Bakinov, I.I. Vaneev, S.R. Gorlovsky, V.I. Eropkin, R.V. Zashinkin and A. S. Koniv, 7th Int. Miner. Process. Congr. Gordon and Breach, New York, NY, 227 (1964).

[9] M.K. Rhodes, 13th Int. Miner. Process. Congr. Warsaw, Poland, (1981).

[10] S.I. Gorlovsky, O.S. Bogdanow (ed.), Flotation Reagents Research, p. 157 (1965).

[11] R. Jin, W.B. Hu and S. Meng, Preprint AIME/SME Annual Meeting. Denver, p. 7 (1987).

[12] Watari, T., Nansai, K., & Nakajima, K. (2020). Review of critical metal dynamics to 2050 for 48 elements. Resources, Conservation and Recycling, 155, 104669. DOI:10.1016/j.resconrec.2019.104669

[13] Thunwall, M., Boldizar, A., Rigdahl, M., Banke, K., Lindström, T., Tufvesson, H., & Högman, S. (2007). Processing and properties of mineral-interfaced cellulose fibre composites. Journal of Applied Polymer Science, 107(2), 918-929. DOI:10.1002/app.26740

[14] Liu, Q., Zhang, Y., & Laskowski, J. S. (2000). The adsorption of polysaccharides onto mineral surfaces: an acid/base interaction. International Journal of Mineral Processing, 60(3-4), 229-245. DOI:10.1016/s0301-7516(00)00018-1

[15] X. Fu, W. Lu, D.D.L. Chung, Ozone treatment of carbon fiber for reinforcing cement, Carbon 36 (9) (1998) 1337-1345.

[16] Y. Xu, D.D.L. Chung, Improving the workability and strength of silica fume concrete by using silane treated silica fume, Cem. Concr. Res. 29 (3) (1999) 451-453.

[17] C. Wang, K.Z. Li, H.J. Li, G.S. Jiao, J. Lu, D.S. Hou, Effect of carbon fiber dispersion on the mechanical properties of carbon fiber-reinforced cement-based composites, Mater. Sci. Eng. A 487 (2008) 52-57

[18] Akbar, A. Y., Lestari, Y., Ramadhan, G., Candra, S. A., & Sugiarti, E. (2014). The Influence of Carboxy Methyl Cellulose (CMC) and Solution pH on Carbon Fiber Dispersion in White Cement Matrix. Applied Mechanics and Materials, 493, 661-665. DOI:10.4028/www.scientific.net/amm.493.661

[19] Klemm, D., Heublein, B., Fink, H.P., Bohn, A.: Cellulose: fascinating biopolymer and sustainable raw material. Angew. Chem. 44, 3358-3393 (2005).

[20] Varshney, V.K., Naithani, S.: Chapter 2: chemical functionalization of cellulose derived from nonconventional sources. In: Kalia, S., Kaith, B.S., Kaur, I. (eds.) Cellulose Fibers: Bio- and Nano-polymer Composites, pp. 43-60. Springer, Berlin (2011)

[21] NEDO (New Energy and Industrial Technology Development Organisation). CWM in Japan. NEDO International Committee; 1997]

[22] Boylu, F., Ateşok, G., & Dinçer, H. (2005). The effect of carboxymethyl cellulose (CMC) on the stability of coal-water slurries. Fuel, 84(2-3), 315-319. DOI:10.1016/j.fuel.2003.12.016.

[23] Mianehrow H, Moghadam MH, Sharif F, Mazinani S: Graphene-oxide stabilization in electrolyte solutions using hydroxyethyl cellulose for drug delivery application. Int J Pharm. 2015 Apr 30;484(1-2):276-282. doi: 10.1016/j.ijpharm.2015.02.069. Epub 2015 Feb 28.

[24] Wang J, Somasundaran P: Mechanisms of ethyl(hydroxyethyl) cellulose-solid interaction: influence of hydrophobic modification. J Colloid Interface Sci. 2006 Jan 15;293(2):322-332. Epub 2005 Aug 2.

[25] Feng, B., Jiao, X., Wang, H., Peng, J., & Yang, G. (2021). Improving the separation of chalcopyrite and galena by surface oxidation using hydroxyethyl cellulose as depressant. Minerals Engineering, 160, 106657. DOI:10.1016/j.mineng.2020.106657

[26] Vasconcelos, H., JMMM de Almeida., Saraiva, C., Jorge, Pedro A S., Coelho, L., 2019. Preliminary study for detection of hydrogen peroxide using a hydroxyethyl cellulose membrane. Proceedings, 15(1)

[27] Zecher, D. and Van Coillie, R. (1992) Cellulose derivatives, in Thickening and Gelling Agents for Food (ed. A. Imeson), Blackie A & P, Glasgow, pp. 40-65.

[28] Walton, A.G. and Blackwell, J. (1973) in Biopolymers, Academic Press, New York, pp.464-74.

[29] Greminger, G.K., Jr and Krumel, K.L. (1980) Alkyl an hydroxyalkylcellulose, in Handbook of Water-soluble Gums and Resins, ch. 3 (ed. R.L. Davidson), McGraw-Hill, New York, pp. 3.1-3.25.

[30] Ramby, B.G. and Rydholm, S.A. (1956) Cellulose and cellulose derivatives, in Polymer Processes, vol. x (ed. C.E. Schildknecht), Wiley Interscience, New York, pp. 351-429.

Chapter 3

Nano-Cellulosic Fibers from Agricultural Wastes

Nozieana Khairuddin, Md. Bazlul Mobin Siddique, Mohammad Sobri Merais, Nurul Husna Che Hamzah and Dayangku Nurshahirah Awang Wahab

Abstract

In recent years, the potential of agricultural wastes has received increasing attention from academia and industry. The aim has been to identify strategies for the conversion of low-value wastes into new materials and other value-added products. Cellulose is a naturally abundant polymer that is readily available in various agricultural wastes. It is a linear polymer consisting of β-D-glucopyranose units (disaccharides) joined by glycosidic β-1,4 bonds. Nanoparticles can be extracted from cellulose fibers using a top-down mechanically or chemically treatment. Cellulose nanomaterials have generated significant interest due to their intrinsic properties such as large surface-to-volume ratios, high tensile strength, stiffness, and flexibility in addition to good dynamic mechanical, electrical, and thermal properties. The use of nanocellulose for reinforcement in matrices improves thermo-mechanical properties, decreases the sensitivity of polymers to water, and preserves biodegradability. The mixing of nanocellulose with polysaccharides improves mechanical properties. Nano-sized cellulose fibers possess unique physical, chemical, and morphological characteristics. Hence, nano-sized cellulose fibers are considered versatile materials for addition to polymers, and application in high gas barriers and packaging materials. Other uses include electronic devices, foods, medicine, cosmetics, and health care. This chapter focuses on the cellulose nanofibers attained from banana, pineapple and corn-based agricultural wastes.

Keywords: fiber, post-harvest, nanocellulose, chemical hydrolysis, enzymatic hydrolysis, agricultural waste

1. Introduction

1.1 Agricultural industry

Agricultural development is one of the most powerful tools used to boost shared prosperity and economic growth as well as to alleviate extreme poverty and feed an estimated population of 9.7 billion people by 2050. In 2018, agriculture accounted for 4% of global gross domestic product (GDP) and over 25% of the GDP of some developing countries. However, current food systems pose threats to human health and the planet since agriculture accounts for 70% of water use, while generating alarming rates of pollution and wastes. According to the World Bank (2021), one-third of the food produced globally is either lost or wasted [1]. **Table 1** shows the GDP of Malaysia by sector from 2019 to 2021 (at constant 2015 prices).

1.2 Crop production

1.2.1 Banana

Banana, scientifically known as *Musaceae*, originated from India to Papua New Guinea, and parts of Southeast Asia [2, 3]. Banana is one of the most popular fruits worldwide that contains various essential nutrients such as potassium, magnesium, Vitamin B6 and niacin that aid in digestion, healthy heart, and weight loss [4]. The rich taste of banana is used in desserts, parfaits, and baked goods. It can be consumed raw or after cooking via frying, boiling, stewing or few other methods [5]. India is the world largest banana and plantain producer. As of 2019, the production of bananas in India was 30.5 million tonnes, which accounts for 26.02% of the global production. The other top global banana and plantain producers include; China, Indonesia, Brazil and Ecuador accounting for 53.94% of the production. It was estimated that the total global production of bananas in 2019 was 117 million tonnes [6]. The main parts of the banana plant consist of the leaves, fruit bunch, suckers, underground stem and male flowers as depicted in **Figure 1a** while b) shows statistical data of the top ten largest banana producing countries and contributors in 2019.

1.2.2 Pineapple

Pineapple (*Ananas comosus*) is a healthy and delicious tropical fruit that originated in South America, where early European explorers named it after a pinecone. It is rich in various minerals and vitamins especially vitamin C and manganese, which are essential for maintaining a healthy immune system and body metabolism. It can be purchased fresh, canned or frozen for a daily diet [8]. The pineapple plant consists of a crown, slip, fruit, leaf, stem and sucker as shown in **Figure 2a**. **Figure 2b** displays the graphical statistics of the worldwide production of pineapple in 2019. As reflected in **Figure 2b**, Costa Rica is the largest producer of pineapple accounting for 3.33 million metric tonnes in 2019. This is followed by the

	Share (%)		Change (%)	
	2020[1]	2019	2020[1]	2021[2]
Services	58.1	6.1	−3.7	7.0
Manufacturing	22.6	3.8	−3.0	7.0
Agriculture	7.4	2.0	−1.2	4.7
Mining	6.9	−2.0	−7.8	4.1
Construction	4.0	0.1	−18.7	13.9
GDP	100	4.3	−4.5	6.5–7.5

[1]*Estimate.*

[2]*Forecast.*

Note: Total may not add up due to rounding and exclusion of import duties.
Source: Department of Statistics and Ministry of Finance, Malaysia (2021).

Table 1.
GDP of Malaysia by sector, 2019–2021 (at constant 2015 prices).

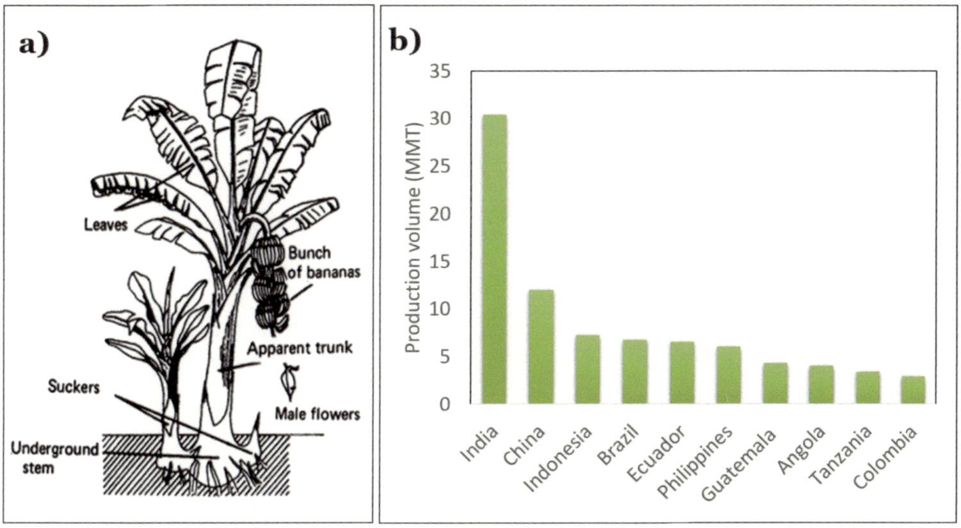

Figure 1.
(a) Parts of the banana plant [7]; (b) Top 10 largest producers of banana in 2019 [6].

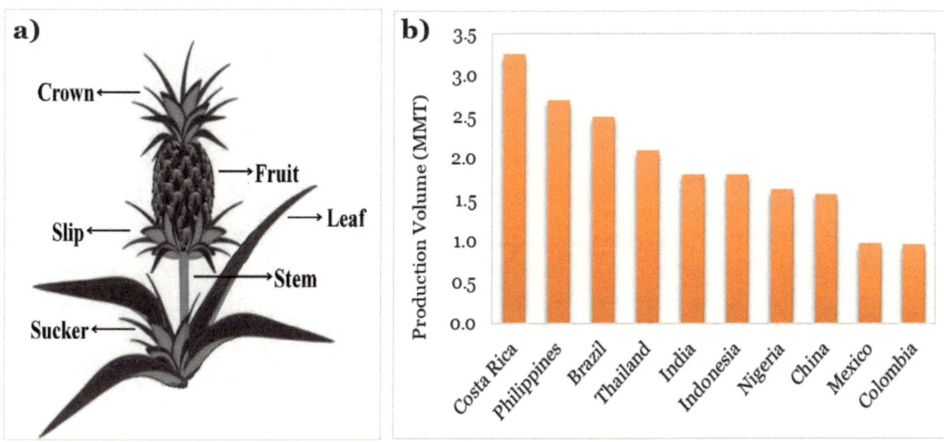

Figure 2.
(a) Parts of pineapple [9]; (b) top 10 largest pineapple producers worldwide [10].

Philippines, Brazil, and Thailand. The total production of pineapples in 2019 was approximately 28.18 million tonnes [10].

1.2.3 Corn

Corn (*Zea mays L.*) is the third largest agricultural crops globally after wheat and rice. Almost 100% grain corns in Malaysia was imported from Brazil, Argentina and other countries. Corn are important source of foods for both humans and animals. There are plenty types of corn planted across the world, which includes dent corn, flint corn, sweet corn, popcorn, flour corn, pod corn and waxy corn. Corn is processed to prepare food and feed ingredients such as breakfast cereals, corn starch, sweeteners (high fructose corn syrup), cooking oil and lysine. Corn is also utilized in manufacturing industrial products such as ethanol and polylactic acid (PLA). The corn stems can be employed to prepare silage for ruminants [11]. While sweet corn has been traditionally popular fruit in the USA, China, and Brazil

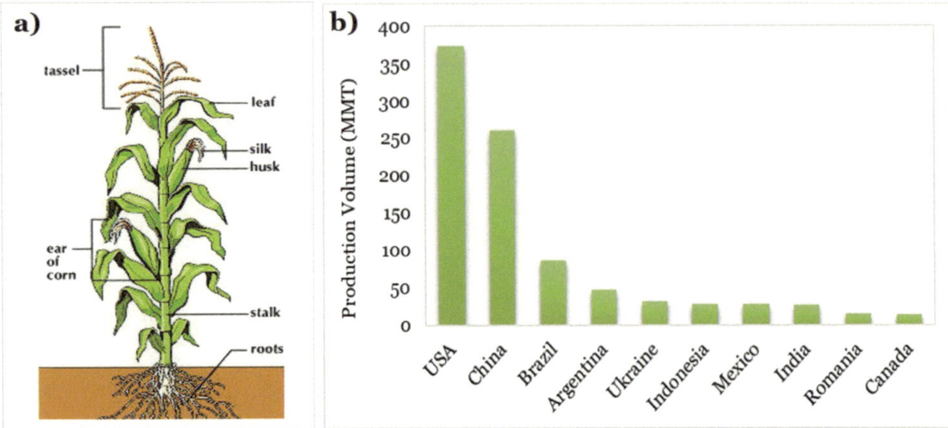

Figure 3.
(a) Parts of corn plant [12]; (b) top 10 largest corn producers worldwide [10].

(**Figure 3**), although it has recently gained popularity in numerous Asian countries including Malaysia [10]. Corn is typically yellow but comes in various other colors such as red, orange, purple, blue, white and black. Whole corn is packed with fiber and contains vitamins B and C and the elements magnesium and potassium that promote gut health and prevent digestive diseases [13].

2. Agricultural wastes

Presently, organic and agro-industrial residues constitutes a great share in overall global wastes as depicted in **Figure 4a** which makes it crucial to effectively manage to exploit it. As biomass feedstocks are abundant, it has high likelihood for future application in biorefinery technologies [14, 16]. It was believed that the biomass wastes in Latin America that has most of the lignocelluloses are mostly acquired from maize, soybean, and sugarcane residues [17].

2.1 Banana waste

Banana residues are not considered among the primary agro-industrial waste to be utilized as biorefinery biomass supply in developing countries. Therefore,

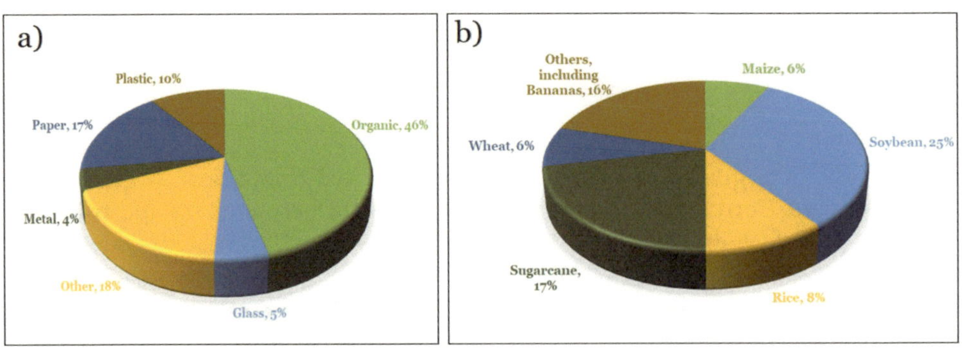

Figure 4.
(a) Composition of global solid waste in 2015 [14]; (b) estimated biomass crop residue flows for Latin America in 2012 [15].

banana waste treatment has become an issue that needs to be resolved [18]. The common banana plant wastes include the pseudostem, floral stalk, banana peels and leaves. Elanthikkal et al. [19] added that banana peel waste is a by-product of processing banana chips and baby foods. The edible part of banana consists of 12 wt% of the whole plant, which indicates the remainder is discarded as agricultural waste thereby causing environmental problems. Chang [20] also mentioned that about 40% of bananas that do not fit standards set by supermarkets are discarded by farmers [2]. The waste produced by a single banana plant can account for nearly 80% of the total plant mass. It is estimated that 220 tonnes of by-products are produced per hectare per year [21], which requires innovative ideas to convert the agricultural wastes into value-added products. **Figure 5a** illustrates a mountain of banana waste being discarded daily as it does not fit the cosmetic look set by the buyers.

2.2 Corn waste

Corn stover is the biggest agricultural waste in the United States (**Figure 6**) due to the high productivity of the farms. Corn stover refers to stalks, leaves and cobs of the remnants of corn harvest. About 100 million dry tonnes of corn stover are collected annually making corn stover not only a renewable and sustainable material but also one of the cheapest and most available residues in the US. Corn cobs serve as a novel source of raw material for the microbial production of fuel alcohols

Figure 5.
(a) Mountains of bananas being discarded daily [20]; (b) waste banana fruit peels [22].

Figure 6.
(a) the process of harvesting corn [23]; (b) corn Stover leftover [24].

and other value-added products [25, 26]. Corn harvest consists of 48% stalk of the total dry mass, 28% leaves, 15% cobs, and 8% husk [27].

2.3 Pineapple waste

The pineapple fruit is typically processed into jam, juice, cordial, vinegar, and food flavoring. Solid or biomass wastes such as core (5–7%), peels (30–35%), a crown (10–15%), and stems are accumulated from the processing or harvesting of the pineapple fruit and plants, as shown in **Figure 7**. In addition, the rough handling of fruits and exposure to extreme environmental conditions during transportation or storage contributes about 55% of product waste [30]. After harvesting, the pineapple waste is burnt or left on the ground, which can cause serious environmental issues. It is estimated that 40—80% of pineapple fruit wastes have high biological oxygen demand (BOD) and chemical oxygen demand (COD) values [31].

2.4 Environmental issues

The poor management of agricultural wastes through practices such as open burning and dumping in landfills causes environmental pollution. For instance, high moisture content of banana waste will eventually produce greenhouse gases if it is not properly managed (e.g. by dumping) thereby resulting in disastrous impacts on the environment. Farmers are known to throw banana tree wastes in rivers, ponds, lakes or in-low lying areas where slow degradation occurs thereby generating methane, other gases, and spreading putrescible odors that affect the nearby ecosystem. The farming practice of open fire burning to eliminate bananas wastes also contributes to severe air pollution issues [21]. Fruit residues also cause severe environmental problems as it accumulates in the agro-industry.

Fruit wastes will eventually lead to serious environmental problems as it accumulates in the agro-industrial yard with zero significance and commercial value. The high costs of transportation and the limited availability of landfills have resulted in the unscrupulous disposal of wastes into the environment. Additionally, the problems of disposing of by-products have worsened due to legal restrictions. Pineapple wastes have high BOD and COD values, which exacerbates management and disposal [32]. Scientists have focused on the co-digestion of pineapple wastes together with several other fruits, legume, manure and slaughterhouse wastes to decrease the volatile solids by 50–65% [33]. The environmental problems of pineapple wastes are evident in the world's largest producer, Costa Rica and other countries where the juice industry discards 50–65% of the residual biomass [34].

Figure 7.
Pineapple residues from: (a) peels [28]; (b) crown [29].

As for the management of corn wastes, the burning of crop residue eliminates the opportunity to improve organic matter content and potentially leads to substantial loss of nutrients. Nearly all nitrogen and at least 75% of the sulfur in plant residues are lost during combustion. A primary tillage practice designed to chop stalks, deep rip, and establish raised beds in a single operation can save considerable time and fuel expense, compared to several conventional tillage operations. Therefore, it is important to identify and characterize the waste produced to minimize environmental pollution by optimizing waste reduction [35].

3. Cellulose nanofibers (CNF)

3.1 Background

The cell wall of plants contains cellulose, hemicelluloses and lignin. The lignin component accounts for 10–25% by dry weight and acts as a binder between cellulose and hemicelluloses. It is lignin that enhances rigidity, durability, and protects the cell wall thereby acting as a binder. The other two major components of the plant cell wall are cellulose and hemicelluloses that comprise 35–50% and 20–35%, respectively, of the dry weight of lignocellulosic biomass. Cellulose is the linear polysaccharide with repeating units of cellobiose (disaccharide *d*-glucose) units linked by β-1,4 linkage. Hence, there are strong intramolecular or intermolecular hydrogen bonds between adjacent glucose units in similar or different chains through the open hydroxyl groups present in glucose monomer units. The hemicellulose in plant cell walls is mainly composed of xylan and glucomannan, which are pentose and hexose monomers linked by short or branched chains. The compact structures of hydrogen bonding are tightly packed networks in cellulose fibers that provide antibacterial properties, toughness, strength, and water or solvent impermeability to the plant cell wall. Nanocellulose is a fiber or crystal-based form of cellulose measuring a few micrometers in length and diameter < 100 nm and typically extracted from plant, animal, mineral and other natural sources. It is biodegradable, lightweight with a density of about 1.6 gm/cc with 10 GPa of high tensile strength similar to cast iron. It also accommodates reactive hydroxyl groups which are ideal for surface functionalization for various applications. The three types of nanocellulose materials are cellulose nanofibers (CNF), cellulose nanocrystals (CNC) and bacterial nanocellulose (BNC) [36]. **Table 2** shows the chemical composition of lignocellulosic materials derived from various agricultural wastes.

3.2 Methods of cellulose nanofibers extraction

Numerous studies on the use of agricultural waste (biomass) as sources of CNF have been widely discussed over decades. The processes for extracting

Source	Composition (%)				Refs.
Lignocellulosic biomass	Cellulose	Hemicellulose	Lignin	Extracts, pectin and waxes	
Corncob	28–34	39–47	21–29	5–12	[37]
Pineapple leaf	34–40	21–25	25–29	8–10	[38]
Banana peel	60–65	6–8	5–10		[39]

Table 2.
Chemical composition of lignocellulosic materials from various agricultural wastes.

nanocelluloses (NCs) from biomass are categorized into two main procedures: biomass pre-treatment before NC extraction and method of isolation of NC [36].

3.2.1 Pre-treatment of biomass

The initial step for NC extraction is the pre-treatment of lignocellulosic biomass to eliminate hemicellulose and lignin. The two common methods used for biomass pre-treatment are alkali treatment and acid-chlorite treatment as described below.

3.2.1.1 Alkaline treatment

Alkaline treatment is a process whereby the biomass is treated with an alkali reagent to eliminate non-cellulosic components such as pectins, hemicelluloses and lignin. The filtrate is then washed with water to neutralize it and the obtained solid contains mostly cellulosic components. The common concentration of NaOH used for this treatment is 8–16% and the mixture is heated to 90–160°C for 1–2 h for the removal of hemicelluloses and some parts of lignin. Treatment is usually carried out using a strong base solution such as sodium hydroxide or potassium hydroxide with varying conditions as tabulated in **Table 3**.

Novel approaches such as steam explosion have successfully removed the lignin fraction while producing purer and structurally intact cellulose. Next, the solubilized components are eliminated by rinsing with deionized water and filtered. The conditions applied may vary depending on the fiber source and its constitution [43].

3.2.1.2 Acid-chlorite treatment (bleaching treatment)

The acid-chlorite treatment is the partial or complete removal of lignin from lignocellulosic biomass by combined treatment with sodium chlorite acidified with glacial acetic acid. This process is commonly known as the bleaching or delignification process, which is performed by mixing distilled water, sodium chloride and acetic acid with lignocellulosic biomass at 70–80°C for 4–12 h. The blend is stirred overnight followed by washing with distilled water until neutral (pH = 7) is achieved. The white residual product characterized as holocellulose lignin-free is

Lignocellulosic source	Reagent	Conditions	Refs.
Banana peel	KOH (5% w/v) Bran to KOH ratio (1:20)	27–30°C, 14 h, mechanical stirring	[40]
Banana pseudostem	H_2O_2 (10% v/v), NaOH (1% w/v)	100 rpm, 60 min Autoclave; 110°C, 40 min	[41]
Banana fibers	NaOH 1 M	80°C, 4 h	[19]
Banana fibers	NaOH (2 wt%)	Steam explosion; 20 lb. (12°C /1 h)	[42]
Banana, Pineapple leaf	NaOH (2 wt%)	30°C, 6 h	[43]
Corn stover	NaOH (4 wt%)	150 rpm, 8o°C, 2 h 2x treatment	[44]
Cornstalk	NaOH (15 wt%)	55°C, 2.5 h, mechanical stirring	[45]
Pineapple leaf	NaOH (5%) Fiber to solution ratio 1:20 (w/v)	Water bath; room temperature, 1 h	[46]

Table 3.
Alkaline treatment of different lignocellulosic fibers.

Lignocellulosic source	Reagent	Conditions	Refs.
Banana peel	$NaClO_2$ (1 %w/v) at pH 5 adjusted with acetic acid	27–30°C, 14 h	[40]
Banana pseudostem	H_2O_2 (10% v/v), NaOH (1% w/v)	100 rpm, 60 min in Autoclave; 110°C, 40 min	[41]
Maize straw	H_2O_2 2% (v/v) and TAED 0.2% (w/v)/ HNO_3 65% (v/v)	48°C, 12 h/120°C, 30 min	[47]
Pineapple leaf	NaOH, acetic acid and 1:3 NaClO solution	25°C, 1 h	[48]
Corn stover	50 g NaCl, 500 ml nanopure water, 50 ml glacial acetic acid	30°C, 24 h, mechanically stirred in a fume hood	[44]
Cornstalk	10 ml ethyl ethanol, 10 wt% NaOH	35°C, 2 h, mechanical stirring	[45]

Table 4.
Bleaching treatment of different lignocellulosic fibers.

then collected and oven-dried. **Table 4** shows the varying conditions and reagents used by other researchers for bleaching treatment.

3.2.2 Methods of nanocellulose isolation

In the last decades, various methods to manufacture CNF from regenerated cellulose and natural raw materials have been developed. These methods include electrospinning [49, 50], biosynthesis [51, 52], and mechanical isolation. Some review papers have been published to compare these methods [53–55]. The conventional needle-based electrospinning techniques for producing CNF deteriorates from the use of toxic solvents, very low productivity (spinning rate: 10 ml/h), the use of high voltage [56] and poor crystallinity. Biosynthesis by bacteria yields nanofibers with narrow diameter distributions, high aspect ratios and high crystallinity. However, it has several constraints such as the requirement for strict and high-cost production, poor reproducibility between the bacteria of different generations, and the complex post-treatment purification procedures. According to Sharma [36], the extraction of nanocellulose may be performed via three different methods; acid hydrolysis, enzymatic hydrolysis or mechanical treatment processes.

3.2.2.1 Acid hydrolysis

Acid hydrolysis is a common method used to isolate NC. The amorphous region of the cellulose fibrils may be hydrolysed using strong acids such as sulfuric acid via the esterification of hydroxyl groups by the sulphate ions. It will subsequently make the crystalline region of cellulose fibers to produce a stable colloidal dispersion of nanocellulose materials in the remaining reaction mixture. Sulfuric acid is the most used acid for the hydrolysis process. Other researchers have also used several mild acids such as formic, acetic, phosphoric and chloric acids for hydrolysis as summarized in **Table 5**. This reaction depends on three primary reaction parameters; reaction time, temperature and acid concentration, which also influences the properties of the nanocellulose product. The washing procedure is normally conducted using cold water followed by centrifugation or using sodium hydroxide until neutral pH is obtained. For instance, Pelissari et al. [40] mixed 1% v/v H_2SO_4 solution with the insoluble residue at 80°C for 1 h to eliminate the mineral traces and hydrolyse amorphous cellulose, providing the nanofibers product. The insoluble residue was then neutralized with alkaline or acid solution (5% KOH or 10% acetic acid) depends on the pH followed by centrifugation (10,000 rpm; 5°C; 20 min) until neutral pH was obtained. The final residue was diluted with deionized water and the suspension was stored at 4°C in a sealed bottle.

3.2.2.2 Enzymatic hydrolysis

Enzymatic hydrolysis is a biological treatment process that utilizes enzymes for the digestion or modification of cellulose fibers to obtain pure cellulose. The common enzymes used for this process include; cellulase, endoglucanase, cellobiohydrolase. The mechanism is complex but the enzymatic action is based on breaking or catalytic linking of H-bonds between the cellulose microfibers. The hydrolysis process is critical to the removal of hemicellulose, which protects cellulose from hydrolysis and the production of monosaccharides from hemicellulose for further fermentation to bioethanol. Cellulases and hemicellulases are structurally related both fundamentally and in relation to their reactant systems that typically function in synergy. Both enzymes are usually required for proficient hydrolysis of assorted lignocellulosic biomasses. The process of enzymatic hydrolysis is carried out under mild condition, but the reaction time required is much longer when compared to the acid hydrolysis process.

3.2.2.3 Mechanical treatment process

Cellulose fibers can be mechanically processed to isolate nanocellulose fibers (NCFs) using various mechanical methods such as ultrasonication, ball milling, and

Lignocellulosic source	Reagent	Conditions	Refs.
Banana peel	H_2SO_4 (1 wt%)	27–30°C, 14 h	[40]
Banana pseudostem	H_2O_2 (10 wt%), NaOH (1 wt%)	100 rpm, 60 min Autoclave; 110°C, 40 min	[41]
Corn cob	H_2SO_4 (64 wt%)	45°C, 90 min	[37]
Cornstalk	30 wt% chloroacetic acid-ethanol solution	75°C, 2.5 h	[45]

Table 5.
Acid hydrolysis treatment of different lignocellulosic fibers.

Lignocellulosic source	Mechanical treatment	Conditions	CNF diameter (nm)	Refs.
Banana peel	High-pressure homogenization	Number of passages: 3, 5 & 7 times 1st stage: 500 bar 2nd stage: 50 bar	11–23	[40]
Banana peel	High-pressure homogenization	Number of passages: 5 times 1st stage: 500 bar 2nd stage: 50 bar	310–508	[58]
Corn stover	High-pressure homogenization	Number of passages: 30 times	5–50	[44]

Table 6.
Multiple mechanical treatments used on various sources of lignocellulosic fibers.

high-pressure homogenization. However, the main limitation of these processes is the high input energy requirements, which require the incorporation of pre-treatment methods to reduce energy consumption. Among the mechanical methods, high-pressure homogenization is also widely applied to isolate CNFs from plants [57]. It is important to note that the diameter of CNFs (which usually varies from around 2 to 20 nm in diameter and up to several microns in length) is largely

Extraction method of NCF	
Sample: Fruit peel (unripe mature green *Musa paradisiaca*) [40]	Sample: Pseudostem [41]
Oxidation prevention method • Samples soaked in potassium metabisulfite. Solution (1% w/v) for 24 h to prevent browning (oxidation) Drying samples in the oven; 60°C, 24 h	Freeze-drying method • Samples were frozen at -50°C for 12 h • Freeze-driedfor48 h
Fine grinding and sieving after drying	
1st alkalinisation treatment • Samples treated with 5% w/v KOH solution (ratio 1:20) under constant stirring at room temperature for 14 h	Alkalinisation and bleaching treatment • Dried samples mixed with deionized water. H_2O_2 (10% v/v) and NaOH (1% w/v) at 100 rpm for 1 h • Mixture placed in an autoclave at 110°C and 0.5 kg/cm^3 for 40 min • Vacuum filtering and rinsing with deionized water three times to remove remaining traces of NaOH
Neutralization reaction • Insoluble residue neutralized to pH 7 with alkaline (5% KOH) or add solution (10% acetic acid)	Neutralization reaction • Using H_2SO_4 to neutralize the suspension (pH 7) • Rinsing with deionized water to remove remnant traces of acid
Bleaching treatment (Delignification) • Using 1% w/v $NaClO_2$ at (pH 5) adjusted with acetic acid • Repeated until sample discolored	Acid hydrolysis treatment • A mixture of deionized water, H_2O_2 and H_2SO_4 (6.25% v/v) with the product suspension was kept under constant stirring at 100 rpm for 1 h • The dispersion was autoclave at 110°C and 0.5 kg/cm^2 for 40 min • Vacuum filtering and rinsing with deionized water was carried out three times
Neutralization reaction	
2nd alkalinisation treatment • Similar conditions as the first alkalinisation method Neutralization reaction	
Acid hydrolysis treatment • Insoluble residue mixed with a 1% v/v H_2So_4 solution at 80°C to eliminate mineral traces and to hydrolyse amorphous cellulose to obtain the required nanofibers	
Neutralization reaction	Neutralization reaction
The final product was diluted with deionized water and centrifuged (10,000 rpm; 5°C; 20 min) and the suspension was stored at 4°C in a sealed container Homogenization (mechanical treatment) • 1st stage; 500 bar • 2nd stage; 50 bar • Aliquot suspensions passed through homogenizer 3,5, or 7 times	Drying of product for 48 h at 40°C

Table 7.
Comparison of CNF extraction method from the banana fruit peel and its pseudostem.

determined by the source. **Table 6** summarizes the various conditions of high-pressure homogenization applied by other researchers.

Tables 7–9 further elaborates the methods used by researchers to obtain NCF from the agricultural wastes of banana, corn and pineapples.

3.3 Applications of CNFs

CNFs possess distinctive properties such as high mechanical strength, tunable surface chemistry, high aspect ratio, crystallinity, barrier properties, non-toxicity and biodegradability. The materials have great potential as a source of green packaging, coatings, fillers and other industrial applications [36]. The application of CNF is described hereafter mainly focusing on the wastes from banana, corn and pineapple.

3.3.1 Nanocomposites

Epoxy-based FRP (Fiber-reinforced Polymer) composites have relatively high thermal stability. Nevertheless, it possesses low crack growth resistance that limits mechanical applications. Studies have confirmed that hybridizing fibers and fillers can significantly enhance the properties of epoxy FRPs [59]. The theory of

Extraction method of NCF	
Sample: Corn stover [44]	Sample: Corn stalk [45]
Sample preparation • A chipper shredder was used to process the stover sample that was field dried. The sample was fed through a mill before grinding with a 6-micron screen, then a 1-micron screen, and finally, a 0.5-micron screen.	Sample preparation • The washed corn stalks were dried in a 60 vacuum drying oven for 6 h, smashed, sifted* through a 100-mesh sieve.
Alkali treatment • The sample was suspended in water at 2%(w/v) and blended using a laboratory blender. The sample was rinsed with water through a 75 µm sieve. It was then resuspended and centrifuged. • The supernatant was removed and 500 ml of 4 wt % NaOH was added. • The suspension was stirred at 150 rpm at 80°C for 2 h. Then, it was washed thoroughly with nano pure water using a 75 µm sieve. The treatment was repeated. • After washing, the sample was centrifuged until a clear supernatant was observed and recovered. • The sample was air-dried at room temperature.	Alkali treatment • Corn stalk was weighed and 15 wt% NaOH solution was added and transferred into a 5 ml 3-necked flask. • This reaction mixture was placed into a wa bath kept at 55°C for 2.5 h with mechanic stirring. • The samples were rinsed few times w deionized water and then dried at 60°C. • The weighed sample, 10 mL ethyl alcohol, a 10 wt% NaOH solution were added to a 500 ml 3-necked flask and stirred continuously 35°C for 2 h
Bleaching/ delignification treatment • The delignification solution was prepared using 50 g of NaCl added to 500 ml of nano pure water and 50 ml of glacial acetic acid. • The sample was added to the delignification solution and mechanically stirred at 30°C for 24 h in a fume hood • The sample was air-dried after thorough washing with water.	Acid hydrolysis treatment • 12 g of 30 wt% chloroacetic acid-ethanol solutions was added to the reactor at 75°C, 2.5 h. • The resulting product was rinsed thorough with deionized water repeatedly and fine dried at 60°C to a constant weight.

Table 8.
Comparison of CNF extraction method from corn Stover and corn stalk.

Extraction method of NCF	
Sample: pineapple peels [34]	Sample: pineapple leaves [46]
Alkalinisation and bleaching treatment • The ground pineapple leaves were incubated in NaOH to remove lignin and hemi cellulose, then bleached in $NaClO_2$ before incubating in HCl to hydrolyse the cellulose into micro-cellulose.	Sample preparation • The fibers were extracted from pineapple leaves bunches by scraping. Afterwards, the fibers were crushed using the RL-L10 MPL crusher to reduce the length. • Sieving process was carried out using an SS304 GMP automatic sieving machine for segregated fibers below 2.00 mm.
Neutralization reaction • After each chemical treatment, the product was thoroughly rinsed with water until a neutral pH was obtained. The solid was recovered by centrifugation at 13000 rpm	Alkali treatment • The short pineapple leaf fiber was soaked in 5 wt % of NaOH solution in a water bath for 1 h at room temperature. The ratio of the fibers and the solution was 1:20 (w/v). • The samples were washed and rinsed ten times using distilled water.
Acid hydrolysis • The sample was hydrolysed with H_2SO_4 to attain nanocellulose. • Two fractions of particles were separated from the acid solution by centrifugation at 2500 rpm • The solid was incubated with a second H_2SO_4 solution between 30 min and 4 h at 55°C • The product was separated from the liquid viscous phase by centrifugation at 2500 rpm and rinsed few times using deionized water	Composite preparation • All raw materials for the preparation of the composite samples including the treated and untreated fibers, tapioca biopolymer and coupling agents were oven-dried at 80°C for 24 h.

Table 9.
Comparison of CNF extraction method from pineapple peels and leaves.

Hybrid polymer composites	Lignocellulosic source	Refs.
Kenaf/phenolic resin	Silane treated pineapple leaf fiber (PALF)	[60]
Kenaf/Polylactic acid	Cornhusk flour	[65]
Woven Kenaf/unsaturated polyester resin	Banana fibers	[66]
Glass/vinyl ester	Pineapple leaf fiber	[67]

Table 10.
Nanocomposites based on cellulose nanofibers.

hybridization offers great opportunities for applications in automotive, aerospace and construction [60, 61]. The incorporation of nanofillers in FPR improves the physical, mechanical, morphological, thermal and dynamic properties of the composites [62]. Hence, CNF has become a viable alternative when compared to carbon nanofillers [63] due to its biodegradability, high aspect ratio and significant mechanical properties [64]. **Table 10** summarizes the previous research works on the developed Kenaf hybrid thermoplastics, thermoset and biopolymer composites for advanced applications.

3.3.2 Packaging

The synthetic plastic wastes generated from the food packaging industry are disastrous to the environment if left unrecycled. Hence, the development of biodegradable films and coatings has been widely studied recently. These biopolymers

Packaging blend	Lignocellulosic source	Refs.
Polyvinyl alcohol/polyacrylic acid/NCF	Banana pseudostem	[69]
Polyvinyl alcohol/NCF	Banana peel	[73]
Banana starch/Glycerol/NCF	Banana peel	[74]
Potato starch/Glycerol/NCF	Pineapple leaf	[75]

Table 11.
Packaging based on CNFs.

are renewable, abundant, cost-effective and non-toxic. Furthermore, the tensile strength and elongation at break are considered the main mechanical properties required to preserve the integrity of the packed food. Since nanocelluloses possess these properties, they are ideal components of biodegradable food packaging [68]. The popular product features of NCF-based packaging required by researchers include a shelf-life extension [69], heat resistance [53], and intelligent packaging [70]. Other properties include transparent packaging [71], ultra-violet screening packaging, and anti-microbial packaging [72]. **Table 11** shows the various NCF-based packaging blends made from lignocellulosic banana, pineapple and corn residues for packaging by researchers.

4. Comparison of bioplastics with petroleum-based plastics

The plastics industry began in the early 1900s when the first synthetic plastic was created by Leo Hendrik Baekeland in the United States [76]. Since the industry began, annual global plastic production has grown extensively from some 1.5 million metric tons in 1950 to 359 billion metric tons in 2018. The cumulative production of plastic has already exceeded eight billion metric tons worldwide, with further increase is expected in the future. The increase in plastic waste has become a global environmental issue. Although recycling has become more widely practiced in recently, many plastic materials produced over the last 70 years has ended up in landfills, and often ends up in water bodies, contributing to the environmental pollution [77].

Although plastics production uses less than 5% of all gasoline, its renewability is usually a source of concern. Separating various forms of petroleum-based recyclable plastics from other solid wastes is a time-consuming and labor-intensive process, hence, only a limited proportion of plastics are recycled. The failure of petroleum-based plastics to biodegrade is also a point of contention among environmentalists [78].

Table 12 show a collection of LCA literature data [79, 80]; each LCA characterizes and compares the environmental impact of various bioplastics (thermoplastic starch (TPS), polylactic acid (PLA), and polyhydroxyalkanoates (PHA) and traditional plastics (high and low density polyethylene, Nylon 6, polyethylene terephthalate (PET), polystyrene (PS), polyvinyl alcohol (PVOH) and polycaprolactone) with an approach cradle to grave.

Overall, the statistics in **Table 12** demonstrates how the manufacture and utilization of bioplastics is more energy efficient and reduces greenhouse gas pollution as compared to traditional plastics. In contrast, bioplastics has a significant environmental effect due to soil acidification and eutrophication, owing to the use of fertilizers and chemicals in the processing of organic raw materials used in bioplastics manufacturing. However, the inclusion of non-biodegradable

Type of Plastic	Energy Requirement (MJ/Kg)	Global Warming (Kg CO_2 eq/Kg)
Non-renewable source		
HDPE	80.0	4.84
LDPE	80.6	5.04
Nylon 6	120.0	7.64
PET	77.0	4.93
PS	87.0	5.98
PVOH	102.0	2.70
PCL	83.0	3.10
Renewable source		
TPS	25.4	1.14
TPS +15% PVOH	24.9	1.73
TPS + 60% PCL	52.3	3.60
PLA	57.0	3.84
PHA	57.0	Not Available

Table 12.
Energy requirement and global warming data for each type of plastic.

copolymers in bioplastics necessitates a substantial rise in energy consumption and CO_2 emissions as compared to bioplastics. Indeed, non-biodegradable copolymers are added to mechanical biopolymers to boost their efficiency, lowering the biodegradable capacity of the resulting material. It is vital to note that the above LCA findings were obtained using incineration with energy recovery as the final provision: this option is not especially beneficial to bioplastics due to their low calorific value.

The study of LCA data consistently has confirmed that bioplastics have certain environmental impact indices that are lower than other conventional plastics, although other indices favor the latter; hence, the need to develop an overall environmental impact index that incorporates and accurately weighs all indices [81].

Author details

Nozieana Khairuddin[1,2*], Md. Bazlul Mobin Siddique[3], Mohammad Sobri Merais[1], Nurul Husna Che Hamzah[1] and Dayangku Nurshahirah Awang Wahab[3]

1 Department of Science and Technology, Faculty of Humanities, Management and Science, Universiti Putra Malaysia Bintulu Sarawak Campus, Bintulu, Sarawak, Malaysia

2 Institut Ekosains Borneo, Universiti Putra Malaysia Bintulu Sarawak Campus, Bintulu, Sarawak, Malaysia

3 Faculty of Engineering, Computing and Science, Swinburne University of Technology Kuching, Kuching, Sarawak, Malaysia

*Address all correspondence to: nozieana@upm.edu.my

References

[1] The World Bank. Agriculture Overview [Internet]. Washington DC: The World Bank; [updated 2020 Sep 30; cited 2021 May 4]. Available from: https://www.worldbank.org/en/topic/agriculture/overview

[2] Arvanitoyannis IS, Mavromatis A. Banana cultivars, cultivation practices, and physicochemical properties. Critical Reviews in Food Science and Nutrition. 2009;**49**(2):113-135.

[3] De Langhe E, Vrydaghs L, De Maret P, Perrier X, Denham T. Why bananas matter: an introduction to the history of banana domestication. Ethnobotany Research and Applications. 2009;**7**: 165-177.

[4] Ware M. Benefits and health risks of bananas [Internet]. Orlando: Medical News Today; 2020 January 13 [cited 2021 May 9]. Available from: https://www.medicalnewstoday.com/articles/271157

[5] Ali S. Top 10 largest banana producing countries in the world [Internet]. Countries Now; 2013 March 22 [cited 2020 May 9]. Available from: https://www.countriesnow.com/top-10-largest-producers-of-bananas/

[6] Knoem*a. Banana*s production quantity [Internet]. Knoema; 2020 Dec 22 [cited 2021 May 10]. Available from: https://knoema.com/atlas/topics/Agriculture/Crops-Production-Quantity-tonnes/Bananas-production

[7] Food and Agricultural Organization of United States. Bananas. FAO Economic and Social Development Series. 1992:**3**;10-25

[8] Raman R. 8 Impressive health benefits of pineapple [Internet]. Healthline; 2018 May 26 [cited 2020 May 10]. Available from: https://www.healthline.com/nutrition/benefits-of-pineapple#TOC_TITLE_HDR_4

[9] Prado, KS, Jacinto, Spinacé, MAS. Cellulose Nanostructures Extracted from Pineapple Fibres. Pineapple Leaf Fibres. 2020:185-234.

[10] FAO. Crops [Internet]. FAOSTAT. 2020 Dec 22 [cited 2021 May 19]. Available from: http://www.fao.org/faostat/en/#data/QC/visualize

[11] Nor NA, Rabu MR, Adnan MA, Rosali MH. An overview of the grain corn industry in Malaysia. FFTC Agricultural Policy Platform (FFTC-AP). Retrieved March. 2019;**10**:2020.

[12] Encyclopædia Britannica. Encyclopædia Britannica, inc.; [cited 2021 May 14]. Available from: https://kids.britannica.com/students/assembly/view/53137

[13] Streits L. Is corn good for you? Nutrition facts and more. Healthline; 2018 Nov 30 [cited 2021 May 11]. Available from: https://www.healthline.com/nutrition/is-corn-good-for-you

[14] Yang X, Choi HS, Park C, Kim SW. Current states and prospects of organic waste utilization for biorefineries. In Renewable and Sustainable Energy Reviews. 2015;**49**:335–349.

[15] Sommer SG, Hamelin L, Olesen JE, Montes F, Jia W, Chen Q, et al. Agricultural waste biomass. Supply Chain Management for Sustainable Food Networks. 2015;67–106.

[16] Redondo-Gómez C, Rodríguez Quesada M, Vallejo Astúa S, Murillo Zamora JP, Lopretti M, Vega-Baudrit JR. Biorefinery of Biomass of Agro-Industrial Banana Waste to Obtain High-Value Biopolymers. Molecules. 2020;**25**(17):3829.

[17] Iakovou E, Bochtis D, Vlachos D, Aidonis D, editors. Supply chain management for sustainable food networks. Hoboken, NJ: John Wiley & Sons; 2016 Mar 22.

[18] Padam BS, Tin HS, Chye FY, Abdullah MI. Banana by-products: an under-utilized renewable food biomass with great potential. Journal of food science and technology. 2014c;**51**(12): 3527-3545.

[19] Elanthikkal S, Gopalakrishnapanicker U, Varghese S, Guthrie JT. Cellulose microfibres produced from banana plant wastes: Isolation and characterization. Carbohydrate Polymers. 2010;**80**(3): 852-859.

[20] Chang C. War on Waste: Craig Reucassel reveals the shocking truth about our bananas [Internet]. Australia: News.com.au; 2017 May 16 [cited 2021 May 7]. Available from: https://www.theengineer.co.uk/banana-waste-converted-biodegradable-plastic/#:~:text=Researchers%20in%20Australia%20have%20devised,%2C%20non%2Dtoxic%2C%20bioplastic.&text=According%20to%20the%20researchers%2C%20the,the%20rest%20discarded%20as%20waste.>.

[21] Shah MP, Reddy GV, Banerjee R, Babu PR, Kothari IL. Microbial degradation of banana waste under solid state bioprocessing using two lignocellulolytic fungi (Phylosticta spp. MPS-001 and Aspergillus spp. MPS-002). Process Biochemistry. 2005 Jan; **40**(1):445-451.

[22] Mugira F. Beyond the Garbage [Internet]. WaterSan Perspective; 2012 Mar 29 [cited 2021 May 11]. Available from: https://waterjournalistsafrica.wordpress.com/2012/03/30/beyond-the-garbage/

[23] Myles. Are We Solving The Wrong Problems With Machine Learning? [Internet]. Everyday Analytics. 2018 Dec 6 [cited 2021 May 17]. Available from: https://everydayanalytics.ca/2018/12/are-we-solving-the-wrong-problems-with-machine-learning.html

[24] Eise J. Study shows potential for growth in biofuels from corn stover [Internet]. Indiana: Phys.org; 2015 Nov 13 [cited 2021 May 8]. Available from: https://phys.org/news/2015-11-potential-growth-biofuels-corn-stover.html

[25] Mann L, Tolbert V, Cushman J. Potential environmental effects of corn (Zea mays L.) stover removal with emphasis on soil organic matter and erosion. Agriculture, ecosystems & environment. 2002;89(3):149-166.

[26] Koundinya V. Corn Stover | Agricultural Marketing Resource Center [Internet]. Agmrc.org. 2021 [cited 2021 May 12]. Available from: https://www.agmrc.org/renewable-energy/feedstocks-biofuels/annual-crops/corn-stover#:~:text=Corn%20stover%20is%20the%20largest,percent%20of%20total%20gasoline%20needs.

[27] Analysis of Corn Stover, Cellulose Content of Corn Stover, Lignin Content of Corn Stover [Internet]. Celignis.com. 2021 [cited 2021 May 12]. Available from: https://www.celignis.com/feedstock.php?value=27

[28] Magoum, I. BENIN: ENABEL to finance five pineapple waste recovery projects [Internet]. Afrik 21; 2020 Nov 25 [cited 21 May 2021]. Available from: https://www.afrik21.africa/en/benin-enabel-to-finance-five-pineapple-waste-recovery-projects/

[29] Rueda M. Plates made from biodegradable scraps grow edible plants [Internet]. Business Insider. 2021 Apr 23 [cited 2021 May 15]. Available from: https://www.businessinsider.com/lifepack-colombia-plates-pineapple-tops-crowns-2020-11

[30] Nunes MC, Emond JP, Rauth M, Dea S, Chau KV. Environmental conditions encountered during typical consumer retail display affect fruit and vegetable quality and waste. Postharvest Biology and Technology. 2009;**51**(2): 232-241.

[31] Ban-Koffi L, Han Y. Alcohol production from pineapple waste. World Journal of Microbiology and Biotechnology. 1990;**6**(3):281-284.

[32] Upadhyay A, Lama JP, Tawata S. Utilization of pineapple waste: a review. Journal of Food Science and Technology Nepal. 2010;**6**:10-18.

[33] Alvarez R, Liden G. Semi-continuous co-digestion of solid slaughterhouse waste, manure, and fruit and vegetable waste. Renewable Energy. 2008;**33**(4):726-734.

[34] Corrales-Ureña YR, Villalobos-Bermúdez C, Pereira R, Camacho M, Estrada E, Argüello-Miranda O, Vega-Baudrit JR. Biogenic silica-based microparticles obtained as a sub-product of the nanocellulose extraction process from pineapple peels. Scientific Reports. 2018;**8**(1):1-9.

[35] Larson E. Post-Harvest Crop Residue Management [Internet]. Mississippi: Mississippi State University; 2011 Sep 26 [2011 Sep 26, cited 2021 May 9]. Available from: https://www.mississippi-crops.com/2011/09/26/post-harvest-crop-residue-management/

[36] Sharma A, Thakur M, Bhattacharya M, Mandal T, Goswami S. Commercial application of cellulose nano-composites–A review. Biotechnology Reports. 2019;**21**:e00316.

[37] Silvério HA, Neto WP, Dantas NO, Pasquini D. Extraction and characterization of cellulose nanocrystals from corncob for application as reinforcing agent in nanocomposites. Industrial Crops and Products. 2013;**44**:427-436.

[38] Dos Santos RM, Neto WP, Silvério HA, Martins DF, Dantas NO, Pasquini D. Cellulose nanocrystals from pineapple leaf, a new approach for the reuse of this agro-waste. Industrial Crops and Products. 2013;**50**:707-714.

[39] Mayadevi MR, Sushama PK, Sandeep S. Effects of in-situ bioconversion of farm residues on growth and quality of banana cv. Nendran in laterite soils of Kerala. Journal of Experimental Biology and Agricultural Sciences. 2017;**5**(3).

[40] Pelissari FM, do Amaral Sobral PJ, Menegalli FC. Isolation and characterization of cellulose nanofibers from banana peels. Cellulose. 2014; **21**(1):417-432.

[41] Flores-Jerónimo G, Silva-Mendoza J, Morales-San Claudio PC, Toxqui-Terán A, Aguilar-Martínez JA, Chávez-Guerrero L. Chemical and Mechanical Properties of Films Made of Cellulose Nanoplatelets and Cellulose Fibers Obtained from Banana Pseudostem. Waste and Biomass Valorization. 2021: 1-9.

[42] De Oliveira FB, Bras J, Pimenta MT, da Silva Curvelo AA, Belgacem MN. Production of cellulose nanocrystals from sugarcane bagasse fibers and pith. Industrial Crops and Products. 2016;**93**: 48-57.

[43] Deepa B, Abraham E, Cherian BM, Bismarck A, Blaker JJ, Pothan LA, Leao AL, De Souza SF, Kottaisamy M. Structure, morphology and thermal characteristics of banana nano fibers obtained by steam explosion. Bioresource Technology. 2011;**102**(2): 1988-1997.

[44] Xu J, Krietemeyer EF, Boddu VM, Liu SX, Liu WC. Production and characterization of cellulose nanofibril

(CNF) from agricultural waste corn stover. Carbohydrate Polymers. 2018; **192**:202-207.

[45] Ma D, Zhu B, Cao B, Wang J, Zhang J. Fabrication of the novel hydrogel based on waste corn stalk for removal of methylene blue dye from aqueous solution. Applied Surface Science. 2017; **422**:944–952.

[46] Jaafar J, Siregar JP, Piah MB, Cionita T, Adnan S, Rihayat T. Influence of selected treatment on tensile properties of short pineapple leaf fiber reinforced tapioca resin biopolymer composites. Journal of Polymers and the Environment. 2018;**26**(11):4271-4281.

[47] Rehman N, de Miranda MI, Rosa SM, Pimentel DM, Nachtigall SM, Bica CI. Cellulose and nanocellulose from maize straw: an insight on the crystal properties. Journal of Polymers and the Environment. 2014;**22**(2):252-259.

[48] Cherian BM, Leão AL, De Souza SF, Thomas S, Pothan LA, Kottaisamy M. Isolation of nanocellulose from pineapple leaf fibres by steam explosion. Carbohydrate Polymers. 2010;**81**(3): 720-725.

[49] Frey MW. Electrospinning cellulose and cellulose derivatives. Polymer Reviews. 2008;**48**(2):378-391.

[50] Kim CW, Kim DS, Kang SY, Marquez M, Joo YL. Structural studies of electrospun cellulose nanofibers. Polymer. 2006;**47**(14):5097-5107.

[51] Brown AJ. XLIII.—On an acetic ferment which forms cellulose. Journal of the Chemical Society, Transactions. 1886;**49**:432-439.

[52] Recouvreux DO, Rambo CR, Berti FV, Carminatti CA, Antonio RV, Porto LM. Novel three-dimensional cocoon-like hydrogels for soft tissue regeneration. Materials Science and Engineering: C. 2011;**31**(2):151-157.

[53] Siró I, Plackett D. Microfibrillated cellulose and new nanocomposite materials: a review. Cellulose. 2010; **17**(3):459-494.

[54] Khalil HA, Bhat AH, Yusra AI. Green composites from sustainable cellulose nanofibrils: A review. Carbohydrate polymers. 2012;**87**(2): 963-979.

[55] Lavoine N, Desloges I, Dufresne A, Bras J. Microfibrillated cellulose–Its barrier properties and applications in cellulosic materials: A review. Carbohydrate polymers. 2012;**90**(2): 735-764.

[56] Daiyong Y. Preparation of nanocellulose. PROGRESS IN CHEMISTRY-BEIJING-. 2007;**19**(10): 1568.

[57] Nakagaito AN, Yano H. The effect of morphological changes from pulp fiber towards nano-scale fibrillated cellulose on the mechanical properties of high-strength plant fiber based composites. Applied Physics A. 2004;**78**(4):547-552.

[58] Tibolla H, Pelissari FM, Martins JT, Vicente AA, Menegalli FC. Cellulose nanofibers produced from banana peel by chemical and mechanical treatments: characterization and cytotoxicity assessment. Food Hydrocolloids. 2018; **75**:192-201.

[59] Saba N, Jawaid M. Epoxy resin based hybrid polymer composites. In Hybrid Polymer Composite materials 2017:57-82.

[60] Asim M, Paridah MT, Saba N, Jawaid M, Alothman OY, Nasir M, et al. Thermal, physical properties and flammability of silane treated kenaf/ pineapple leaf fibres phenolic hybrid composites. Composite Structures. 2018;**202**:1330-1338.

[61] Saba N, Paridah MT, Abdan K, Ibrahim NA. Dynamic mechanical

properties of oil palm nano filler/kenaf/ epoxy hybrid nanocomposites. Construction and Building Materials. 2016;**124**:133-138.

[62] Saba N, Paridah MT, Abdan K, Ibrahim NA. Effect of oil palm nano filler on mechanical and morphological properties of kenaf reinforced epoxy composites. Construction and Building Materials. 2016;**123**:15-26.

[63] Saba N, Jawaid M. Recent advances in nanocellulose-based polymer nanocomposites. Cellulose-Reinforced Nanofibre Composites. 2017:89-112.

[64] Li J, Zhou M, Cheng G, Cheng F, Lin Y, Zhu PX. Fabrication and characterization of starch-based nanocomposites reinforced with montmorillonite and cellulose nanofibers. Carbohydrate Polymers. 2019;**210**:429-436.

[65] Kwon HJ, Sunthornvarabhas J, Park JW, Lee JH, Kim HJ, Piyachomkwan K, Sriroth K, Cho D. Tensile properties of kenaf fiber and corn husk flour reinforced poly (lactic acid) hybrid bio-composites: Role of aspect ratio of natural fibers. Composites Part B: Engineering. 2014;**56**:232-237.

[66] Alavudeen A, Rajini N, Karthikeyan S, Thiruchitrambalam M, Venkateshwaren N. Mechanical properties of banana/kenaf fiber-reinforced hybrid polyester composites: Effect of woven fabric and random orientation. Materials & Design (1980-2015). 2015;**66**:246-57.

[67] Zin MH, Abdan K, Mazlan N, Zainudin ES, Liew KE, Norizan MN. Automated spray up process for Pineapple Leaf Fibre hybrid biocomposites. Composites Part B: Engineering. 2019;**177**:107306.

[68] Blanco A, Monte MC, Campano C, Balea A, Merayo N, Negro C. Nanocellulose for industrial use: cellulose nanofibers (CNF), cellulose nanocrystals (CNC), and bacterial cellulose (BC). In Handbook of Nanomaterials for Industrial Applications. 2018:74-126.

[69] Ponni P, Subramanian KS, Janavi GJ, Subramanian J. Synthesis of Nano-film from Nanofibrillated Cellulose of Banana Pseudostem (Musa spp.) to Extend the Shelf life of Tomato. BioResources. 2020;**15**(2):2882-2905.

[70] Khalil HA, Davoudpour Y, Saurabh CK, Hossain MS, Adnan AS, Dungani R, et al. A review on nanocellulosic fibres as new material for sustainable packaging: Process and applications. Renewable and Sustainable Energy Reviews. 2016;**64**:823-836.

[71] Boufi S, González I, Delgado-Aguilar M, Tarrès Q, Pèlach MÀ, Mutjé P. Nanofibrillated cellulose as an additive in papermaking process: A review. Carbohydrate polymers. 2016; **154**:151-166.

[72] Osong SH, Norgren S, Engstrand P. Processing of wood-based microfibrillated cellulose and nanofibrillated cellulose, and applications relating to papermaking: a review. Cellulose. 2016;**23**(1):93-123.

[73] Balavairavan B, Saravanakumar SS. Characterization of Ecofriendly Poly (Vinyl Alcohol) and Green Banana Peel Filler (GBPF) Reinforced Bio-Films. Journal of Polymers and the Environment. 2021:1-6.

[74] Pelissari FM, Andrade-Mahecha MM, do Amaral Sobral PJ, Menegalli FC. Nanocomposites based on banana starch reinforced with cellulose nanofibers isolated from banana peels. Journal of Colloid and Interface Science. 2017;**505**: 154-167.

[75] Balakrishnan P, Sreekala MS, Kunaver M, Huskić M, Thomas S. Morphology, transport characteristics

and viscoelastic polymer chain confinement in nanocomposites based on thermoplastic potato starch and cellulose nanofibers from pineapple leaf. Carbohydrate Polymers. 2017;**169**: 176-188.

[76] Crespy D, Bozonnet M, Meier M. 100 Years of Bakelite, the Material of a 1000 Uses. Angewandte Chemie International Edition. 2008;**47**(18): 3322-3328.

[77] Tiseo I. Topic: Global plastic waste [Internet]. Statista. 2021 Sep 4 [cited 2021 May 19]. Available from: https://www.statista.com/topics/5401/global-plastic-waste/

[78] Kuruppalil Z. Green plastics: an emerging alternative for petroleum-based plastics. Int J Eng Res Innov. 2011; **3**(1):59-64.

[79] Patel, M. 2002. Life cycle assessment of synthetic and biological polyesters. Proceedings of theInternational Symposium on Biological Polyesters, Munster, Germany, September 22–26.

[80] Patel, M. 2005. Starch-Based Technology. In: Handbook of Biodegradable Polymers. Shrewsbury, MA: Rapra Technology Limited, pp. 431–466.

[81] Gironi F, Piemonte V. Bioplastics and petroleum-based plastics: strengths and weaknesses. Energy sources, part a: recovery, utilization, and environmental effects. 2011;**33**(21):1949-1959.

Chapter 4

Deinking of Mixed Office Waste (MOW) Paper Using Enzymes

G. Ramiro Escudero, González P. Jeovani and Perez S. Rosa Elena

Abstract

Deinking of Mixed Office Waste [MOW] paper was carried out by using a flotation device and adding enzymes as defibrillators. Employing the computational simulation of the molecular coupling between the cellulase enzyme obtained from *Trichoderma* sp. and cellulose, the enzyme-cellulose molecular complex and the main amino acids endo-β-1,4-D-glucanase of this molecule, responsible for the hydrolysis of cellulose, were obtained. Three of these amino acids were used for deinking. The deinked cellulose fibers were evaluated according to the standards of the paper industry (ISO whiteness [brightness], factor of reflectance, opacity, and tonality) to determine the efficiency of deinking. The experimental results show that the best optical characteristics of the sheets of deinked paper with amino acids are those where a mixture of aspartate, glutamate, and asparagine was applied, instead of their individual dosage. In addition to the aforesaid, the use of enzymes instead of the traditionally used reagent, such as sodium hydroxide, avoids the contamination of wastewater; additionally, the operation of the column is carried out easily, taking into consideration that the pH of the system goes from alkaline to neutral.

Keywords: paper deinking, flotation column, cellulose, aspartate, glutamate, asparagine

1. Introduction

The recycling of paper is a necessary option for the care of the environment and to comply with the environmental regulations established by governments toward the pulp and paper industry, and specifically, the industry that processes recycled paper fibers to obtain pulp free of ink and that includes in its process the passivation of colloidal particles [1]. Traditional deinking paper (e.g., office paper) is industrially "cleaned" by adding sodium hydroxide (defibrillator) and adding chemical reagents to improve the quality of recycled paper, such as whiteness and mechanical strength [2, 3]. This traditional process undoubtedly damages the environment and specifically urban water reservoirs due to the additional chemical agents contained in the recycled pulps [4–6].

The deinking process exposes the deinking equipment to highly alkaline conditions (pH greater than 10) because the addition of chemical reagents, so ideal conditions in conditions close to neutral pH are preferred, regardless of the chemical

IntechOpen

reagents used to defribillate the cellulose, capture ink particles, and improve optical and mechanical properties of the recycled paper [5–9].

The conditioning of the cellulose pulp is essential to ensure the detachment and capture of the ink particles. The conditioning of the pulp by adding enzymes will promote the disintegration of the paper and the detachment of the ink, as the enzyme is introduced into the cellulose-ink interface. The catalytic action of the enzymes reduces the activation energy of the reactions that occur in the system, even at low concentrations of the reagent, which implies a low contamination of the conditioning medium. Like any chemical reagent, enzymes function within specific pH ranges [10–12]; for example, enzymes called cellulases hydrolyze cellulose fibers and enhance the detachment of ink particles [13–15], which will be available for capture, for example, through the use of devices that generate air dispersions, such as the case of flotation devices [6, 16].

This research work presents the results of the conditioning of recycled office paper pulp (Mixed Office Waste [MOW] type) using enzymes and capturing the ink particles through a flotation column, equipment that is commonly used for mineral processing. The variables that were quantified in the deinked paper to establish the feasibility of the use of enzymes are whiteness, reflectance, opacity, and tonality.

2. Materials and methods

2.1 Enzymatic deinking

Mixed Office Waste (MOW)-type paper was used. The sheets were cut and disintegrated in an industrial blender without chemical reagents. The amino acids such as aspartate, glutamate, and asparagine (Sigma-Aldrich) were added (individually and mixed of all three), as hydrolytic reagents to promote ink detachment.

Once the cellulose pulp had been conditioned for 30 minutes by adding the corresponding chemical reagents, it was fed to a laboratory flotation column made of transparent acrylic tubes. The device (gas disperser) responsible for the generation of bubbles was installed at the external base of the column (venturi-type disperser). The consistency of the pulp fed to the column was set at 4.0%. The experimental conditions are shown in **Table 1**.

In all experiments, pine oil was added as a surfactant to fix and maintain the surface tension at 0.63 N/m. **Figure 1** shows the experimental setup for paper deinking.

Variables	1	2	3	4	5	6	7	8
Amino acid	**ASP**	**ASP**	**GLU**	**GLU**	**ASN**	**ASN**	**ASP:GLU: ASN**	**ASP:GLU: ASN**
Concentration, %	0.1	0.2	0.1	0.2	0.1	0.2	0.1:0.1:0.1	0.2:0.2:0.2
Consistency, %	4	4	4	4	4	4	4	4
pH	7	7	7	7	7	7	7	7
Temperature, °C	25	25	25	25	25	25	25	25
Surfactant addition, ppm	100	100	100	100	100	100	100	100

ASP, aspartate; GLU, glutamate; ASN, asparagine. The % is calculated based on the dry paper mass.

Table 1.
Experimental conditions for deinking with amino acids.

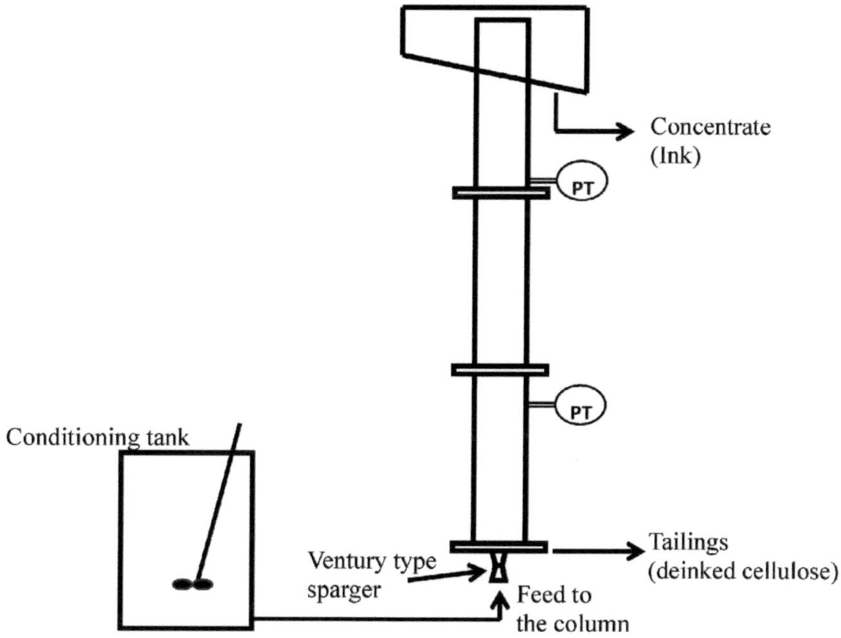

Figure 1.
Experimental setup for deinking of recycled paper. PT = pressure transducer devices.

The gas holdup in the column was estimated through the pressure drop values from pressure transducers installed in two points along the column. The mean bubble diameter of a swarm was calculated by solving the Drift Flux Analysis model [17].

2.2 Evaluation of deinking

The TAPPI T-205-OM-81 standard was applied to the samples collected in the tailings stream (deinked fiber), for each experimental variation of **Table 1**. According to the mentioned standard, the quantified characteristics are whiteness, reflectance, opacity, and tonality.

2.3 Molecular coupling

The method of genetic algorithms coupled to local search or Lamarckian (AUTODOCK 3.0) was used to identify the geometry between the cellulose fragments (ligand) and the cellulase enzyme (substrate).

3. Results and discussion

3.1 Molecular coupling between cellulose and enzyme cellulase

The molecular model representing five cellulose monomer units was taken as a basis. **Figure 2** shows the theoretical model on which this analysis is based.

From the database of the simulation program AutoDockTools, the cellulase enzyme is extracted into its endo-β-1,4-D-glucanase (rigid substrate), as shown in **Figure 3**.

The results from the simulation were analyzed using the AutoDockTools suite with the option of analyzing the conformational space of the ligand substrate. The initial geometry of the studied complex is shown in **Figure 4**.

Figure 2.
Molecular model of cellulose (ligand) of five monomeric units of cellulose.

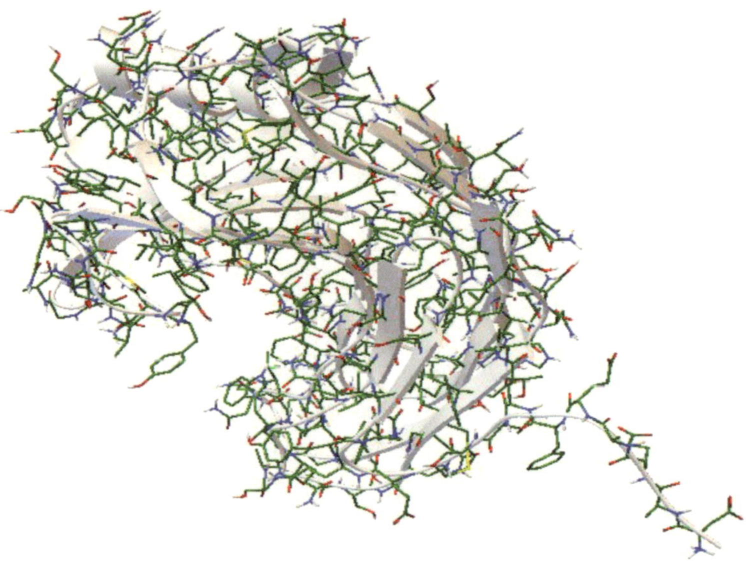

Figure 3.
Schematic representation of the endo-β-1,4-D-glucanase enzyme molecule.

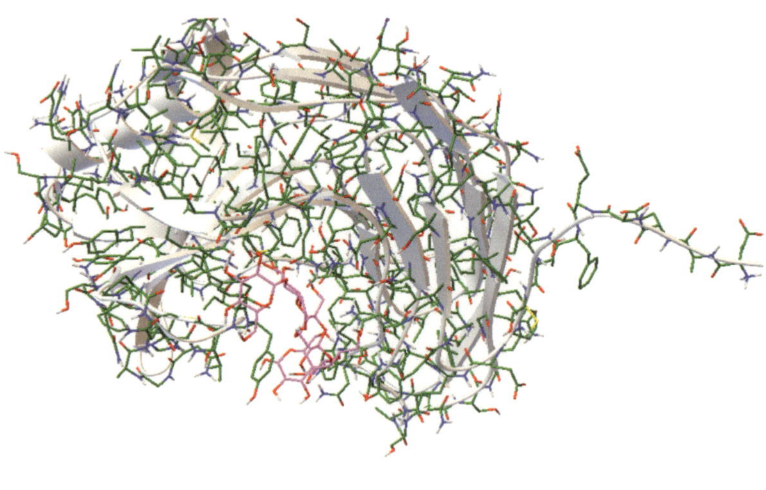

Figure 4.
Initial geometry of the complex enzyme-cellulose.

From the results, the crystallographic coordinates of the crystallized ligand are taken as a reference, grouping the 100 conformations evaluated and obtained in clusters. Each docking is validated after running the same evaluation five times. In the conformational search process, the Lamarckian genetic algorithm (LGA) is applied. The Intelec parameter is activated, so that the internal electrostatic of the ligand affects the energy of docking (docked energy) that guides the optimization process, but not to the binding energy of the ligand (binding free energy). The algorithm analyzes 100 possible poses of the cellulose in the enzyme, adopting the flexibility property for the cellulose molecule, so as not to rule out this interaction phenomenon between ligand and substrate.

The 100 poses analyzed are grouped into the so-called clusters in such a way that they are ordered from the highest to lowest docking energy (binding free energy). This binding free energy is the ligand-substrate coupling with the highest affinity as it has the most negative or lowest bond free energy. It is worth emphasizing that AUTODOCK returns two energies: the so-called docked energy, used to guide the docking, and the binding free energy, the latter allows us to define the ligand-substrate coupling with the highest affinity, that is to say, with the more negative bond free energy. For the case of endo-β-1,4-D-glucanase and cellulose, the pose with the highest affinity drawn the following parameters: binding energy −0.57 kcal/mol, inhibition constant 379.8 mM, intermolecular energy −9.52 kcal/mol, internal energy −14 Kcal/mol, and torsional energy of 8.95 Kcal/mol.

Figure 5 shows the amino acids with the most negative enzyme-cellulose bonds, according to the model: glutamate (GLU, A:207), asparagine (ASN, A:27), tyrosine (TYR, A:118), aspartate (ASP, A:106), tryptophan (TRP, A:29), and phenylalanine (PHE, A:108), whereas **Table 2** shows, according to Morrison and Neilson, the structures of amino acids that interact with cellulose molecules of the endo-β-1,4-D-glucanase.

The experimental results are shown in **Table 3**. A relevant characteristic to determine the efficiency of the deinking process is whiteness, and it is determined through the measurement of the brightness. Brightness ISO is determined using standard sensitivity spectrum centered at a wavelength 457 nm according to ISO 2470 or TAPPI T525 standard and is represented in percentage points. In tests 1 and 2, an ISO whiteness of 85 and 86%, respectively, with 445.2 and 412.5 ppm of black points were obtained. For the L * tonality, there are values of 93.05 in test 1, and 94.25, in test 2. The a * and b * tones manifested with practically equal values, 1.6

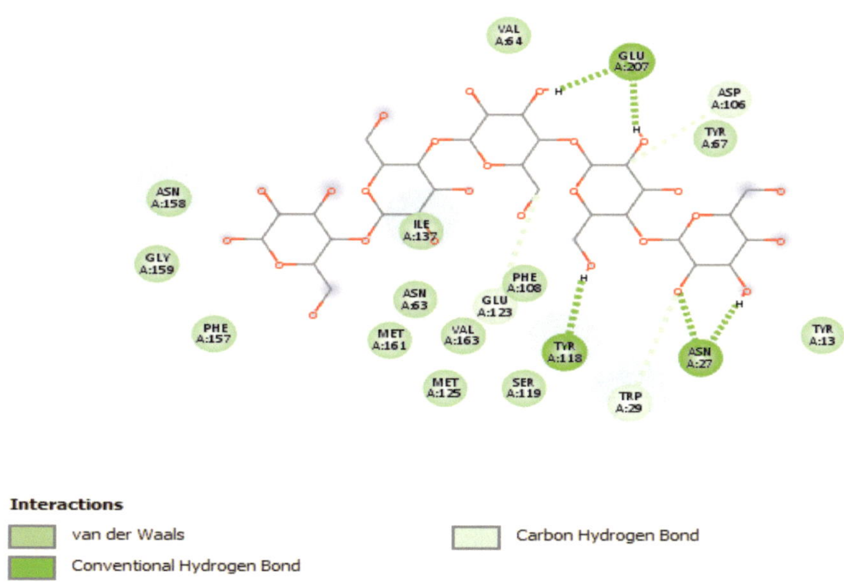

Figure 5.
Linking amino acids in the best enzyme-cellulose according to the modeling.

Amino acid	Structure
Glutamate (GLU)	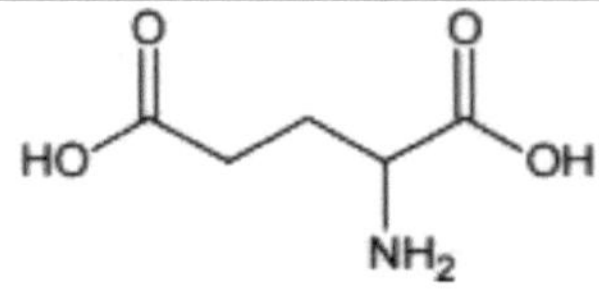
Asparagine (ASN)	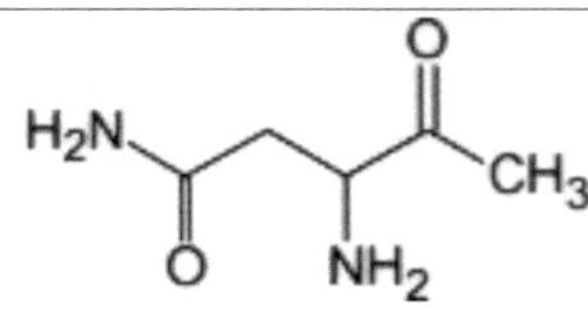
Tyrosine (TYR)	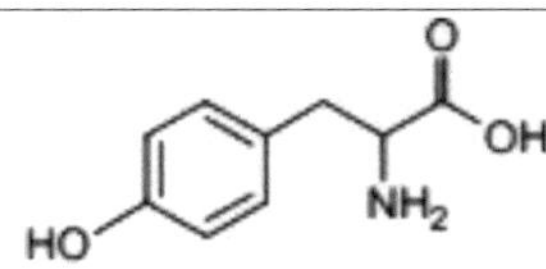
Aspartate (ASP)	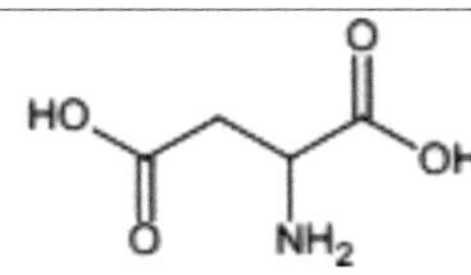
Tryptophan (TRP)	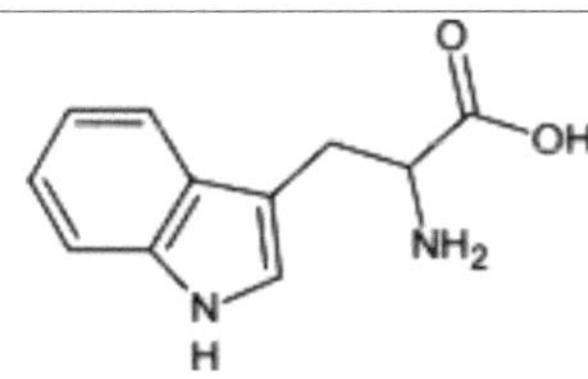
Phenylalanine (PHE)	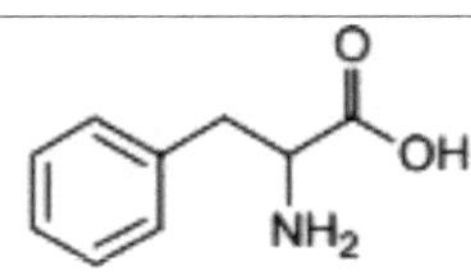

Table 2.
Structures of the amino acids that interact with the cellulose molecule [18].

Variables	1	2	3	4	5	6	7	8
	ASP	ASP	GLU	GLU	ASN	ASN	ASP:GLU: ASN	ASP:GLU: ASN
%	0.1	0.2	0.1	0.2	0.1	0.2	0.1:0.1:0.1	0.2:0.2:0.2
Brightness (ISO), (%)	85	86	84.9	86.9	81.5	85.8	89.1	90.8
Black points, (ppm)	445.2	412.5	455.3	377.1	550.8	423.3	333.0	303.4
L*	93.05	94.25	93.8	94.44	91.6	94.6	94.7	95.3
a*	1.6	1.6	1.67	1.27	1.5	1.3	1.65	1.7
b*	−3.3	−3.36	−2.55	−2.1	−2.8	−2.5	−4.3	−4.52
Reflectance (ISO, %)	41.5	42.7	39.57	47.15	40.2	39.32	35.9	34.7
Opacity (%)	88.3	89.5	88.9	89.9	88	89.5	89.6	89.9

Table 3.
Results of the optical characteristics were measured to the deinked sheets utilizing the use of amino acids.

and − 3.3 for both cases; the same occurs in reflectance and opacity, with very close values, 41.5 and 42.7% of reflectance, and 88.3 and 89.5%, of opacity.

For glutamate, increasing the dose from 0.1 to 0.2% gives an increase in the whiteness of 2%, that is, from 84.9% it rises to 86.9%. Black spots are a direct reflection of inking quality and flotation effectiveness, and this number also decreases with increasing GLU dosage from 455.3 to 377.1 ppm. The tonality variables L *, a *, and b *, such as opacity, change slightly, and reflectance, on the other hand, is modified to a great extent from 39.57 to 47.15%, as the dose of the amino acid in question increases.

When treating the MOW paper with asparagine 0.1%, important changes are noted concerning the tests with aspartate and glutamate. When 0.1% ASN is dosed, the ISO whiteness obtained is 81.5%, the black points 550.8 ppm, and the L * of 91.6, these being much lower than those obtained in the rest of the amino acids. For this case, the variables a *, b *, reflectance, and opacity are practically similar to what was observed with the rest of the amino acids.

Mixing the three amino acids (test 7), adding 0.1% of each depending on the weight of the dry paper to be inked, a result of whiteness much higher than that obtained in the other tests is obtained. The resulting whiteness was 89.1% and the black points 333.0 ppm. The variable b * changes to a lesser extent to −4.3. The parameters L *, a *, and opacity remain above the average of those obtained in the first tests; however, the reflectance decreases considerably to 35.9%, this being inversely proportional to the whiteness.

In test 8, MOW paper deinked with 0.2% of each amino acid obtained the best quality results, ISO whiteness of 90.8% and black points of 303.4 ppm, values much higher than expected, taking into account that the MOW paper is not being treated with any chemical bleach and is only working with three process stages: pulp, chemical treatment, and flotation. The tonality parameters achieved: L * = 95.3, a * = 1.7, and b * = −4.52, oscillate in the values obtained in the previous tests and fall within the expected, according to the values established by the standard, which are L *, from 92.5 to 95.5; a * from 1.3 to 2.3; b * from −9 to −8, reflectance from 32 to 40% and minimum accepted opacity of 87%.

The variable b * is controlled with a pigment from aniline, and the negative of its value indicates the color shift toward blue. The addition of such pigment in the manufacture of paper is directly proportional to the negativity of this parameter, that is, the higher the aniline, the more negative the b *. In deinking, these pigments responsible for the color are detached from the vicinity of the fiber and removed by flotation, so a decrease in this parameter is to be expected after deinking. Opacity is required on all printing papers; it should be sufficient to prevent the printing on the reverse side of the paper from negatively affecting the appearance of a print, so the higher the opacity, the higher the print quality. The result in this test was 89.9%.

The optical properties evaluated in this work are higher than those obtained under similar experimental conditions by using NaOH and the enzyme cellulase *Thricodema* Sp., as defibrillators [19].

3.2 Morphology of the deinked fibers

Through scanning electronic microscope (SEM), there is possible to appreciate the morphology of the deinked fibers, to appreciate the degree of fibrillation, because its intensity affects the mechanical properties of the fiber. In the micrographs of the pulp deinked with aspartate, glutamate, asparagine (**Figures 6–10**), and a mixture of them, it is observed that the depolymerization induced by amino acids does not represent excessive fibrillation that affects the internal morphology of the cellulose fiber and therefore its mechanical properties. No particles of toner

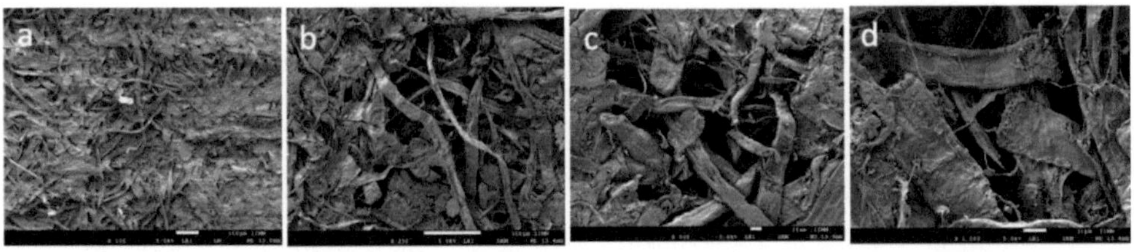

Figure 6.
Micrographs of the deinked fibers. Aspartate case. Test 4: (a) 100X, (b) 250X, (c) 500X, (d) 1000X.

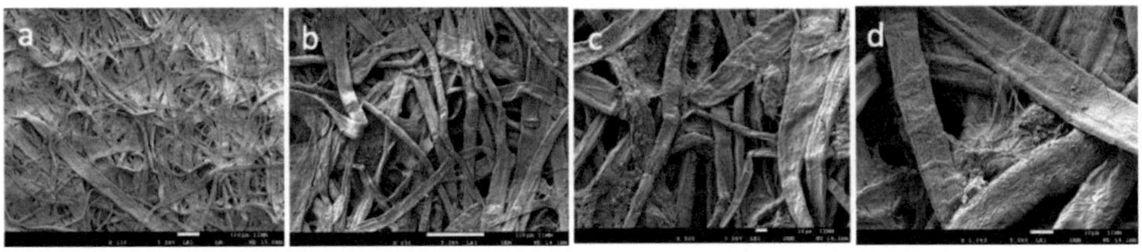

Figure 7.
Micrographs of the deinked fibers. Glutamate case. Test 5: (a) 100X, (b) 250X, (c) 500X, (d) 1000X.

Figure 8.
Micrographs of the deinked fibers. Aspargine case. Test 6: (a) 100X, (b) 250X, (c) 500X, (d) 1000X.

Figure 9.
Micrographs of the deinked fibers. 0.1:0.1:0.1 mixture. Test 7: (a) 100X, (b) 250X, (c) 500X, (d) 1000X.

Figure 10.
Micrographs of the deinked fibers. 0.2:0.2:0.2 mixture. Test 8: (a) 100X, (b) 250X, (c) 500X, (d) 1000X.

or printing ink are seen, only small agglomerations, possibly of starch, precipitated calcium carbonate, and/or residues of other additives typical of the manufacture of paper.

Figure 10 shows the micrograph of fibers from test 8, which obtained the best results of the optical characteristics, from it can be seen that the morphological structure of the fiber retains its robustness, and some fibers look flatter than others, as a result of forming the sheets of paper; however, there is no fibrillation or excessive fine formation. The fragmentation of inks is another phenomenon that occurs in deinking processes, which is desirable up to a certain point so that the particles formed from these acquire the appropriate size to be floated more efficiently. When this happens, papers with superior brightness and whiteness are obtained, without considerable repercussions on the mechanical properties of the deinked fiber.

4. Conclusions

In the present work, through the mathematical simulation and analysis of the molecular coupling of the enzyme to cellulose, by the docking technique, the enzymes with the best coupling between the enzyme-cellulose complexes were aspartate, glutamate, and asparagine.

From the paper deinking experiments using these enzymes, the chemical mixture of the three amino acids results in the highest values of optical properties of the paper sheets formed with the deinked fibers: ISO whiteness or brightness (90.8), reflectance (34.7), opacity (89.9), and tonality (L^* = 95,3, a^* = 1.7, and b^* = −4.52), being these values higher than those obtained through conventional methods.

It is worth mentioning the importance of the flotation column during the deinking, due to the fact of the number of bubbles available for capturing the ink particles.

Author details

G. Ramiro Escudero[1*], González P. Jeovani[1] and Perez S. Rosa Elena[2]

1 Institute of Research in Metallurgy and Materials, Universidad Michoacana de San Nicolás de Hidalgo, Morelia, Michoacán, México

2 Faculty of Chemistry Pharmacobiology, Universidad Michoacana de San Nicolás de Hidalgo, Morelia, Michoacán, México

*Address all correspondence to: ramiro1963@gmail.com

References

[1] Monte MC, Concepción S, Mónica BS, Ángeles NA, Carlos TM. Improving deposition tester to study adherent deposits in papermaking. Chemical Engineering Research and Design. 2012;**90**:1491-1499

[2] Ferguson L. Deinking chemistry: Part 1. Tappi Journal. 1992;**75**:75-83

[3] Carré B, Magnin L, Galland G, Vernac Y. Deinking difficulties related to ink formulation, printing process, and type of paper. Tappi Journal. 2000;**83**:60-92

[4] Beneventi D, Carré B. The mechanisms of flotation deinking and the role of fillers. Progress in Paper Recycling. 2000;**9**:77-85

[5] Rosencrance S. Non-sulfite neutral deinking for recycled fiber. Pulp and Paper Canada. 2007;**108**:15-18

[6] Alzate GHH, Botero AEC, Velázquez JJA, Quintana MGC, Dovale SAM, Charla LV. Study of the enzymatic/neutral deinking process of waste photocopy paper. O'PAPEL. 2013;**74**:61-65

[7] Jobbins J, Heise O. Neutral deinking of mixed office waste in a closed-loop process. In: Recycling Symposium Proceedings. Tappi Press; 1996. pp. 1-6

[8] Haynes R, Merza J. Global application of sulfite-based neutral deinking technology. In: Engineering, Pulping, and Environmental Conference. Tappi Press; 2005. pp. 1-22

[9] Bobu E, Ciolacu F. Deinkability of mixed prints: Alkaline vs neutral deinking. Progress in Paper Recycling. 2008;**18**:23-31

[10] Castro R, Bothig S. Study of Some Variables of the Process of Deinking. Uruguay Technological Lab (LATU); Laboratorio Tecnológico de Uruguay 1994. pp. 1-14

[11] Wodward J, Stephan L, Koran L, Wong K, Saddler J. Enzymatic separation of high-quality uninked pulp fibers from recycled newspaper. Biotechnology 1994;**12**:905-908

[12] Gill R. Advances in the use of fiber modification enzymes in papermaking. In: XXI Tecnicelpa/vi ciadicyp, Portugal. Tappi Press; Vol. 58; 2000

[13] Pala H, Mota M, Gama FM. Enzymatic versus chemical deinking of non-impact ink printed paper. Journal of Biotechnology. 2004;**108**:79-89

[14] Pathak P, Bhardwaj NK, Singh AK. Optimization of chemical and enzymatic deinking of photocopier waste paper. BioResources 2011;**6**(1):447-463

[15] Halperin I, Wolfson H, Nussinov R. Principles of docking: An overview of search algorithms and a guide to scoring functions. Proteins 2002;**47**:409-443

[16] Lavecchia A. Machine-learning approaches in drug discovery: Methods and applications. Drug Discovery Today 2015;**20**:318-331

[17] Escudero R, Finch J. Determining equivalent pore diameter for rigid porous spargers. The Canadian Journal of Chemical Engineering. 2000;**78**:785-792

[18] Morrison & Neilson. Quimica Orgánica, 5ta Ed. Persoon Addison Wesley Longman; 1998

[19] Jeovani González P, Ramiro Escudero G. Deinking of recycled paper by coupling of the enzyme endo-β-1, 4-D-glucanasa and the cellulose, and by using a laboratory flotation column. MRS Advances. 2020;**5**:1-10. DOI:10.1557/adv.2020.390

Chapter 5

Bacterial Cellulose: Multipurpose Biodegradable Robust Nanomaterial

Agata Kołodziejczyk

Abstract

One of actual global problem is clothes and packing materials biodegradability leading to tremendous water contamination. In order to develop ecologically friendly, game-changing in global industry fabric production, we propose a concept to implement kombucha. Kombucha is a symbiotic bacteria and yeast multispecies consortium producing the most abundant polymer on Earth - bacterial cellulose. There are many advantages of bacterial cellulose that are widely used in medicine, material science, food industry and waste management. Unfortunately: long time of bacterial cellulose polymerisation process, lack of its control, diversity in biological composition, finally, acidic smell and disturbances of kombucha growth - all this issues limit the interest of kombucha use to replace easy-accessible and widely applied synthetic materials. In this chapter will be described a revolutionary concept to develop practical and sustainable use of bacterial cellulose as natural alternative for synthetic materials, particularly for a synthetic fabrics and plastics replacement. The optimal cultivation conditions and examples of bacterial cellulose in applications for daily life will be explained.

Keywords: nanocellulose, hydrogel, bio-fabric, design

1. Introduction

One of the most abundant polymers on Earth is cellulose, which is the dominant constituent of plants. It is widely used in daily life in the form of wood and cotton [1–3]. Unfortunately, cellulose obtained from plants has low crystallinity and is contaminated by other polymers like lignins, pectins or hemicelluloses [4]. Such cellulose requires purification processes, which are complex and need use of toxic chemicals, energy and water. Taking the above into account, a more simple and environmentally friendly solution may be use of bacteria-derived cellulose (BC) [5]. BC is a biopolymer synthesised in the fermentation process by various bacteria of genera *Gluconobacter* or *Agrobacterium* [6–9]. The bacteria-derived cellulose is possible the strongest naturally synthesised biological material, which characterises exceptional physicochemical properties, such as high purity [10], crystallinity, water holding capacity, thermal and radiation resistance, mechanical properties, specific surface area, elasticity, relatively high mechanical strength in the wet state, hydrophilicity and excellent biocompatibility [11–15] BC is a very convenient material when it comes to modifications of its applicability since cultivation method determines different shapes, properties and transformations [7]. Among several methods of producing bacterial cellulose, the most simple and ecological one is the use of a symbiotic consortium of bacteria and yeast popularly known as kombucha or SCOBY. Life kombucha cultures are easy to access on the global market and easy to cultivate. Kombucha does not require laboratory conditions, complex growing media or sophisticated cultivation and processing equipment. The optimal growth is obtained on the commonly used sweetened black tea infusion. Each person in the world can grow bacterial cellulose at home without special training. Obtained material is ready to use as an ecological substitute of plastics, storage bags, bandages, and even clothes.

2. Kombucha biological structure

Kombucha microbial consortium (KMC), is a fermented tea and non-alcoholic beverage prepared with water, tea, sugar and kombucha culture (Symbiotic Consortium Of Bacteria and Yeasts known as SCOBY or tea fungus). The Kombucha community is not found in nature. It has been originated by mankind and known already around 220 B.C in Manchuria, northeast China, when it was appreciated for its detoxifying and energising properties [16]. Kombucha microbial consortium is an example of the advanced mutualistic interactions between representatives of two kingdoms of living organisms: bacteria and yeasts [17] (**Figure 1**). The core bacterial community within kombucha brewing is dominated by bacteria *Komagataeibacter, Acetobacter,* and *Gluconobacter* [19]. It may contain lactic acid bacteria [20, 21], but the most remarkable genera characterised in kombucha microbial consortium are *Bacteroides* and *Prevotella* known as dominant human gut microbiota species [22]. Recent metagenomics screening predicted the presence of opportunistic bacteria like *Bacillus, Pseudomonas* etc. [23, 24], bacteriophages and even yeast viruses [25]. Yeast fraction changes in genera composition depending on geographical origin, and it consists of representatives of *Zygosaccharomyces, Brettanomyces, Schizosaccharomyces, Saccharomyces*, and *Pichia* [19]. The biofilm, as a three-dimensional microbial hydrogel settlement, supports an evolutionary stable social cooperation between its inhabitants, in a way that is analogous to tissues in multicellular organisms. The biofilm may also optimise oxygen concentrations for the microbial microcolonies, which are stratified in the cellulosic matrix, as well as

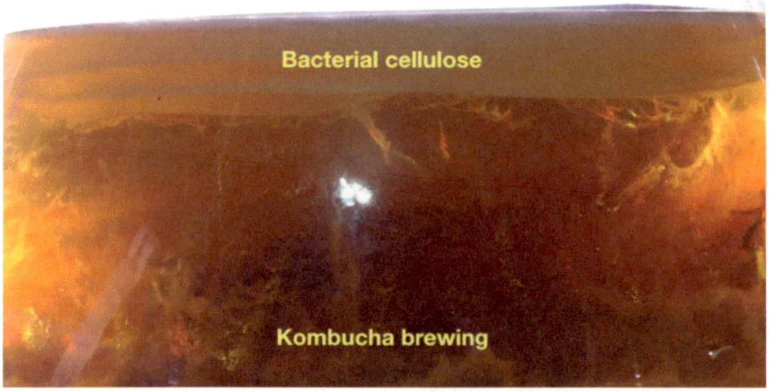

Figure 1.
Kombucha microbial consortium consists of multiple symbiotic species of bacteria and yeasts. On the surface acetic bacteria synthesise the most abundant polymer on our planet - cellulose (modified from [18]).

protect the community from environmental stressors, such as UV radiation [26]. The capability of kombucha microbial consortium to generate and tolerate acidic conditions, metabolise ethanol, and produce organic acids, protects the system from invasion by competitor microbes. The bacterially-produced cellulose-based pellicle biofilm may also provide protection from cheaters by inhibiting the diffusion of their extracellular metabolites.

3. Kombucha chemical composition

Kombucha is diverse not only considering the number of living species inside this micro-ecosystem. Mutualistic interactions between all game players implicate synthesis and transformation of multiple chemical components. Similarly as inside the living cell, many parallel processes can be distinguished. Chemical composition of kombucha beverage depends on cultivation substrate, time and temperature of fermentation process, oxygen tension, microorganisms present in the culture, and the applied method of analysis [27].

Chemical analysis of kombucha revealed *organic acids*: acetic, gluconic, glucuronic, citric, L-lactic, malic, tartaric, malonic, oxalic, succinic, pyruvic, usnic; *sugars:* sucrose, glucose, fructose, and cellulose; *vitamins* B1, B2, B6, B12,K_2 and C; 14 amino acids; biogenic amines; purines; pigments; lipids; proteins; hydrolytic enzymes; ethanol; polyols; antibiotics; carbon dioxide; tea polyphenols; minerals Cu, Fe, Mn, Ni, and Zn; anions; insufficiently known products of yeast and bacterial metabolites; rutin and kombucha-specific theobromine [28–30].

Yeasts and bacteria are involved in metabolic activities that utilise substrates in different complementary ways. Yeasts hydrolyse sucrose via enzyme invertase into monosaccharides and produce ethanol mainly from fructose via glycolysis [31–33]. Bacteria use these fermentation products to synthesise organic acids. Gluconic acid is synthesised from glucose while acetic acid derives from ethanol. Bacterial cellulose fibrils are synthesised by acetic acid bacteria and secreted out of cells as linear polysaccharide polymer, where D-glucose units are linked by β-1,4-glycosidic linkages, similar to plant-derived cellulose. BC generates structural hydrogel with interconnected ribbons of around 100 μm in length and 100 nm diameter composed of a three-dimensional nanofibrous network, which is free of both lignin and pectin. Such a peculiar supramolecular structure engineered by nature makes BC stable and robust. In static conditions, the surface of the kombucha brewing is covered by multiple interconnected reticular pellicles of bacterial cellulose. This structure is a natural barrier to protect consortiums from external environments [24]. In dynamic conditions cellulose forms irregular sphere-like particles [25, 26]. Several species of acidic bacteria are responsible for bacterial cellulose formation. Some of them, for example *Komagataeibacter xylinus,* synthesise cellulose polymers, while others have potential in both producing cellulose and fixating nitrogen [32].

Natural and biodegradable kombucha-derived bacterial cellulose has two main drawbacks: brown colour and unpleasant acidic smell [34, 35]. The brown colour derives from melanoidins, which appear in the Maillard reaction between amino acids and reducing sugars. Acidic smell of the material is due to the presence of hard to remove fermentation products, mainly carboxylic acids [36]. Mentioned problems can be reduced by application of purification methods to remove microorganisms with their metabolites from BC, and to bleach the unwanted colour. The most popular method is the use of alkaline treatment. During such a procedure, increasing concentrations from 0.5M to 5M of NaOH solutions are used [1, 3, 4, 37, 38]. The alkaline purification method requires, however, the use of significant amounts of water and neutralizers to obtain materials with neutral pH. Akkus et al. [39] examined the effect of BC pretreatment in a mixture of polyethylene oxide and NaOH on degradation of the material that could be used as a soft tissue replacement. Another substance that may be applied for this purpose is sodium dodecyl sulfate [40]. Ecological method to reduce the brown colour of BC is replacing tea breweries with algae-based cultivation medium [18, 33]. KMC cultures set on naturally basic *Chlorella* extract must be reduced to pH = 6. Bacterial nanocellulose produced in green solutions is white, delicate and glossy (**Figure 2**). Other types of algae tested in order to optimise KMC growth in space are extremophilic microorganisms such as endolythic *Galdieria sulphuraria* and *Cyanidioschyzon merolae*. It is important to note that adding a new organism to the consortium takes time and sometimes it is not possible for KMC to accept a new partner. Alternatively, yeasts can be modified with genetic engineering tools. Described examples indicate that experimenting with application of various pre- and post-treatments brings new diversified physicochemical properties of the cellulose material. At the end of these considerations, the author proposes a very simple method to reduce the brown colour and acidic smell of KBC, which can be applied at home. The method is based on boiling bacterial cellulose for 15 minutes in the solution of any household cleaning, which contains a bleach.

Another disadvantage of KBC is the fact that in dry state it resembles paper. It is brittle and its mechanical properties are poor. Therefore the final product should be coated supplementing the cellulose's network with a glycerol to provide water retention. In this simple way a polymer HydroGel Bacterial Cellulose is obtained, in which both water and glycerol provide lubrication at the molecular level [18, 33]. This treatment significantly improves mechanical properties of the material. Furthermore, such material surfaces could be additionally sealed with stearic acid to improve properties of natural fabrics. Such a wax-like additive which is spread on and adsorbed by the surface reduces the unwanted effects associated with a short-time contact of the material with water (**Figure 3**). The BC reveals ultrastructural differences in the surface of kombucha cellulose material in the three stages of its processing: raw, lyophilised and coated. The largest difference in scanning electron microscopy imaging was revealed in morphology of bacteria and polymerization patterns. After

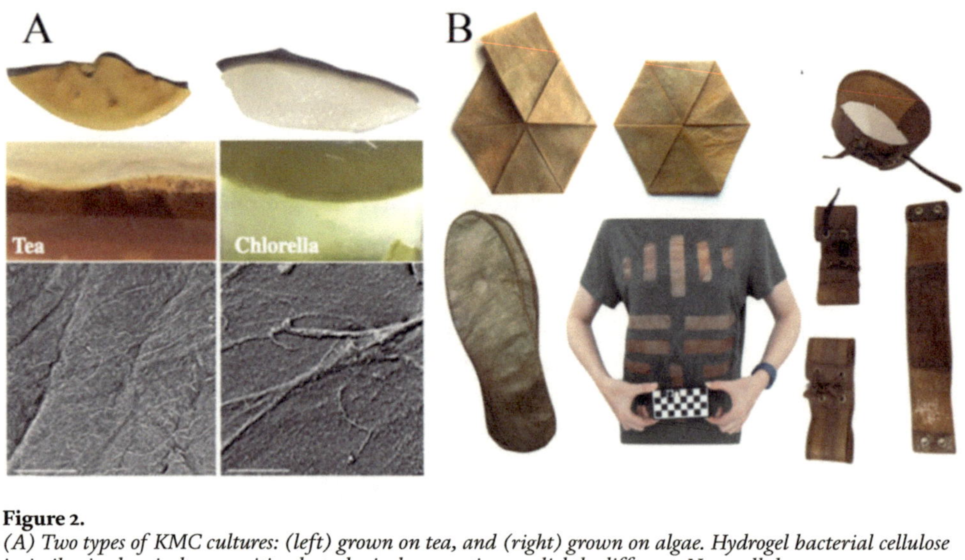

Figure 2.
(A) Two types of KMC cultures: (left) grown on tea, and (right) grown on algae. Hydrogel bacterial cellulose is similar in chemical composition but physical properties are slightly different. Nanocellulose grown on algae infusion is much softer and more prone to damage. Scanning electron microscope images reveal the ultrastructure of nanocellulose filaments. Scale bar 50 microns. (B) KMC's material absorbs odours. It can be easily cleaned and sterilised. It folds gently to very compact volumes. Folding does not implicate changes in structure of the material. Kombucha's cellulose does not generate allergic responses in direct contact with human skin. It makes skin softer and visibly healthier [18, 33].

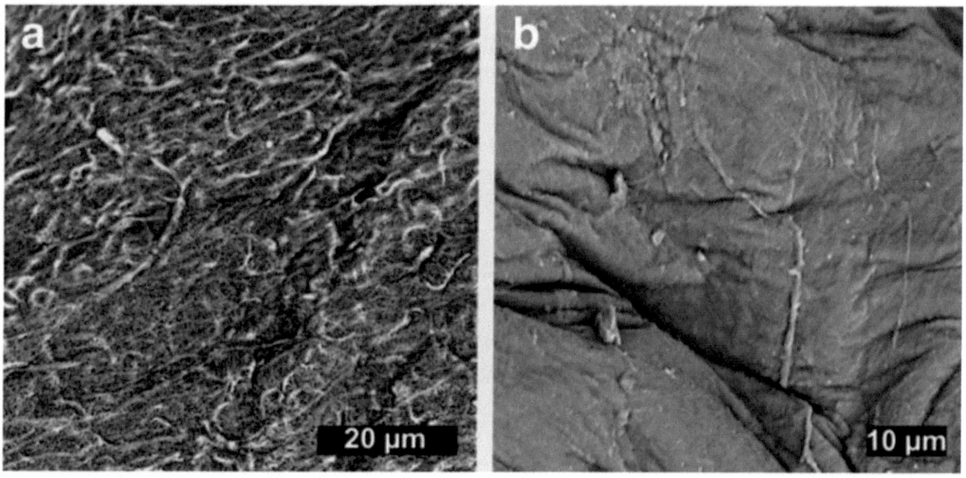

Figure 3.
Scanning electron microscopy images revealing the ultrastructure of kombucha material. (a) A view of a large fragment of the pellicle with yeast (green arrow) and bacteria (red arrow) that are embedded in the cellulose fibres (blue arrows). (b) A close-up of cellulose fibres (blue arrows) and a bacteria cell (red arrow).

sterilization of kombucha samples, no bacteria producing cellulose were visible. Microorganisms were predominantly seen as covered with multiple, distinctive cellulose fibers. After adding the coating solution, the surface of kombucha material has smoothed significantly.

4. Bacterial cellulose

Bacterial cellulose is a very interesting material for the manufacture. First, it may be produced from completely renewable and reusable sources according to the principles of waste-free technologies. Secondly, BC has higher quality for applications because of multiple advantages compared to cellulose produced by plants. It has higher thermal stability, thinner fibrils and higher tensile strength. The highly porous structure of its nanonetwork and controlled shape makes this type of cellulose the source of new ultra-light and stable nanomaterials and cellulose-based nanocomposites. When not needed, BC can easily be degraded with no ecological impact. The material is fully biodegradable, and can be reused for example in soil remediation processes.

There are two types of bacterial cellulose: raw cellulose and processed pure material. Each type of cellulose has several applications in food and package sectors, biomedicine and industry.

The most critical stage of cellulose production is its growth. The BC grows from 3 weeks up to two months depending on cultivation conditions. The procedure of synthesising the almost ready-to-use product from the raw material is very simple. To obtain the most mechanically resistant and optimally thick kombucha material for further processing, 1-2 cm thick raw bacterial cellulose must be grown. The next step is to gently remove the water from the roar hydrogel. It can be done using several methods, such as lyophilisation, drying in the oven, or using mechanical forces. The final thickness of kombucha material is about 0.1mm-1.5mm. Cellulose should be grown in a sterile dish with desired form. For example, to obtain A4 format of a raw pellet of bacterial cellulose, the material should grow in a box of this format. It is important to note, that it should be a minimum 5 cm height of the kombucha brewery level in the dish in order to obtain the best quality of the cellulose. If there is a lower volume of the kombucha solution, the cellulose will not develop properly. The container with kombucha brewing should be separated from the external environment to prevent bio contamination. It should be always covered with a lid allowing air to flow.

Raw cellulose resembles mammal skin. It has high biocompatibility, antimicrobial properties, water holding capacity and valuable metabolites. It is used as wound dressing for skin regeneration and healing, especially after bites of insects. BC exceptional bioaffinity promotes the development of biomedical products for tissue-engineered scaffolds, wound-dressing materials, dental implants, artificial blood vessels and nerves, surgical mesh, bone fillings, heart valve, meniscus, artificial cartilages, etc. [41]. Raw cellulose finds application in the food and package sector in the form of cellulose puree, fat replacer, artificial meat and food package with living biosensors.

Depending on the processing method, different types of cellulose may be produced. The most simple processing method is using cellulose for fabrics and clothes in the form of vegan leather. After drying the raw cellulose and removing the acidic smell, the surface should be coated with one or two layers of glycerin. After drying, the material is ready to use. In order to make the material resistant to water, many types of heterophobic coatings can be applied. One of the simplest solutions and ecological ways is the use of stearic acid to obtain a "wax canvas". Processed and purified bacterial cellulose is applied in production of hydrophilic films for packaging, eye lenses, artificial scaffolds, bandages, clothes, and even fuel cells. When bacterial cellulose is modified with conductive compounds, such as metal nanoparticles or graphite, it naturally becomes conductive for electricity and, therefore, promising in developing materials for electrical applications [42]. KMC living cellulose-based materials can sense and respond to their environment.

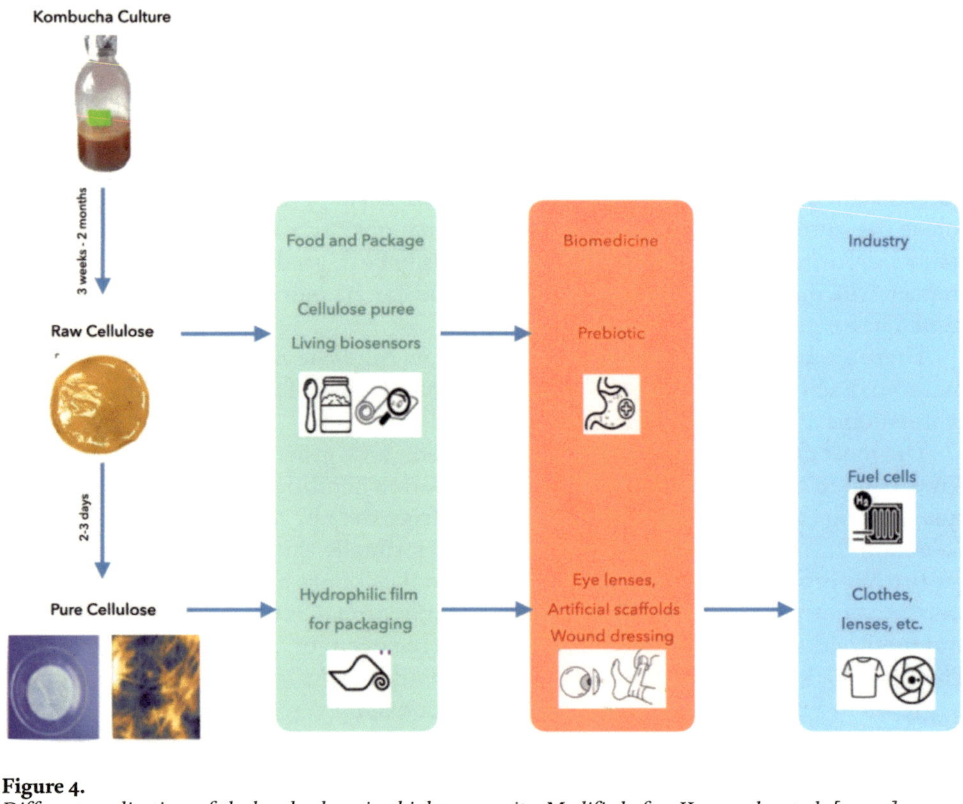

Figure 4.
Different applications of the kombucha microbial community. Modified after Kozyrovska et al. [45–47].

Synthetic biology methods provide tools to create hybrid cellulose materials with predictable novel or enhanced characteristics matching the special purposes of the developing settlements, which cannot be achieved by conventional tools. For example, engineered to secrete enzymes into bacterial cellulose, yeast strains can be taught to sense and respond to chemical and optical stimuli. This means that the modified KMC can be biosensors for many purposes [43]. Moreover, another especially potential use will be intelligent packaging for foods with short shelf-life, as well as biosensors of microbial contamination and mycotoxins.

Growing interest in bacterial cellulose makes it a multipurpose biodegradable robust nanomaterial. The worldwide market for bacterial cellulose is valued at 324.5 million USD in 2020 and is expected to increase put rill 785.1 minion USD in 2026 [44]. Interestingly, since economic and environmental issues will be critical in the early beginning of extraterrestrial settlements, kombucha is considered to be used in space as a sustainable biobased product providing biodegradable and re-usable goods and materials (**Figure 4**).

5. Zero-waste philosophy

Biofabrication of cellulose for industrial purposes based on the kombucha microbial consortium is an elegant example of eco-friendly and zero-waste production. All substrates and side products may be reused and applied in diverse ways. Residues of bacterial cellulose can be used as animal food or in formation of protosoil in various types of agricultural innovative systems including aquaponics, hydroponics and urban gardens. KMC may be stored for many years without need of any type of intervention and can be reused for cultivation any time is needed. Very sour kombucha solution can be used as vinegar, balms, or disinfectants. *Komagataeibacter* cells, as a by-product of the BC production, may be used as antioxidants, to prevent skin dryness.

6. Space application

Space is a challenging environment. Astronauts take all necessary clothes for the time of their mission. They cannot reuse them because there are no washing machines on board spaceships. The price for each kilogram lifted into space ranges from $10 000 to $25 000, which means that clothes in space are very expensive considering long-term missions [48]. There were many discussions about solving the clothing problem in space. One idea is to use bacterial cellulose. KMC-derived nanocellulose may be easily and ecologically processed, even in harsh space conditions [18, 33]. Recent studies revealed that bacterial cellulose retains robustness after 18th months of exposure on the International Space Station. Observed cellulose polymer integrity in exposed samples was not significantly changed. Only after a long-term exposure experiment, the mechanical properties of the newly synthesised cellulose were slightly changed compared to ground control BC pellicles. 2.5 years after the exposure experiment, the kombucha microbial consortium did not return to the initial composition. Among cellulose-producing species, komagataeibacter show the most significant potential of survival in extraterrestrial conditions [49, 50]. This observation indicates the need to modify the bacterial consortium to be more resistant to stressors. Genetic modification of the appropriate candidates for cellulose biofabrication meets increasing attention not only for extraterrestrial conditions but also for earthly use (**Figure 5**).

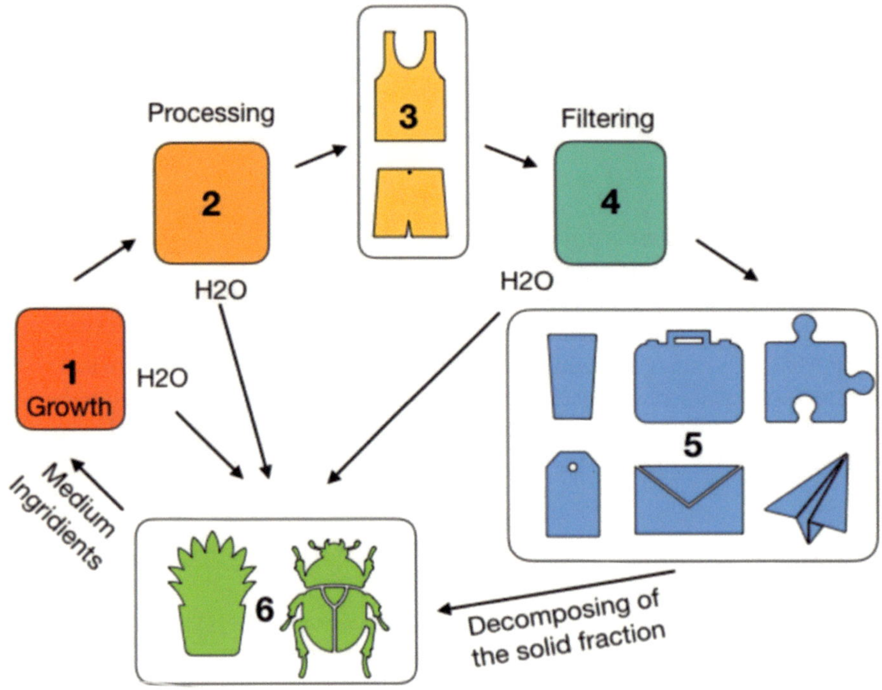

Figure 5.
The summary diagram presents a cycle of multiple processing stages of kombucha cellulose. After bacterial cellulose growth in the kombucha brewery (1), the material is processed by lyophilisation, autoclaving and coating with water-resistant medium (2). Kombucha soft material is used for clothes (3). Used clothes are transformed into filters (4). After drying, the cellulose can be used for various applications as paper, storage boxes, labels, kitchen dishes (5). Finally, bacterial cellulose is digested and decomposed in animal breedings and plant cultivations (6). Converted biomass is then used to prepare a growing medium for the kombucha brewery that encloses the cycle.

7. Summary

Kombucha Microbial Consortium is adapted to live with humans in artificial environments. Several experiments with this microbial community revealed an endless range of plasticity toward desired applications. Despite its health-promoting properties it produces one of the strongest natural biodegradable materials on our planet. A significant advantage of KMC is the zero-waste production in systems ranging all scales of cellulose biofabrication. Its organization as a micro ecosystem provides strong advantages over most microorganisms of biotech value: resistance to contamination, ease of cultivation, and high versatility. Systems and synthetic biology could be used to enhance KMC's biotechnological features but the pure beauty with this multi species community is that everyone can use it at home without sophisticated science. Each person can produce healthy probiotic drinks and eco-fabrics for daily life. Plastic packages and bags can be replaced by home-made bacterial cellulose.

Author details

Agata Kołodziejczyk
Scientific Projects, Analog Astronaut Training Center, Rzeszów, Poland

*Address all correspondence to: agata@astronaut.center

References

[1] Dima SO, Panaitescu DM, Orban C et al (2017) Bacterial nanocellulose from side-streams of kombucha beverages production: preparation and physical-chemical properties. Polymers 9:374(1)–374(2)

[2] Keshk SMAS (2014) Bacterial cellulose production and its industrial applications. J Bioprocess Biotech 4:1000150(1)–1000150(10)

[3] Sederavičiūtė F, Bekampienė P, Domskienė J (2019) Effect of pretreatment procedure on properties of Kombucha fermented bacterial cellulose membrane. Polym Test 78:105941 (1)–105941(7)

[4] Lima HLS, Nascimento ES, Andrade FK, Brigida AIS, Borges MF, Cassales AR, Muniz CR, de Souza Filho MSM, Morais JPS, de Rosa MF (2017) Bacterial cellulose production by Komagataeibacter hansenii ATCC 23769 using sisal juice: an agroindustry waste. Braz J Chem Eng 34:671-680

[5] Brown AJ (1886) An acetic ferment which forms cellulose. J Chem Soc Trans 49:432-439

[6] El-Saied H, Basta AH, Gobran RH (2004) Research progress in friendly environmental technology for the production of cellulose products (bacterial cellulose and its application). Polym-Plast Technol Eng 43:797-820

[7] Gao M, Li J, Bao Z, Hu M, Nian R, Feng D, An D, Li X, Xian M, Zhang H (2019) A natural in situ fabrication method of functional bacterial cellulose using a microorganism. Nat Commun

[8] Matthysse AG, Thomas DL, White AR (1995) Mechanism of cellulose synthesis in Agrobacterium tumefaciens. J Bacteriology 177:1076-1081

[9] Bandyopadhyay S, Saha N, Saha P (2018) Characterization of bacterial cellulose produced using media containing waste apple juice. Appl Biochem Biotechnol 54:649-657

[10] Reiniati I, Hrymak AN, Margaritis A (2017) Recent developments in the production and applications of bacterial cellulose fibers and nanocrystals. Crit Rev Biotechnol 37:510-524

[11] Abitbol T, Rivkin A, Cao Y, Nevo Y, Abraham E, Ben-Shalom T, Lapidot S, Shoseyov O (2016) Nanocellulose, a tiny fiber with huge applications. Curr Opin Biotech 39:76-88

[12] Campano C, Balea Martin A, Blanco A, Negro C (2015) Enhancement of the fermentation process and properties of bacterial cellulose: a review. Cellulose 23:57-91

[13] Habibi Y, Lucia LA, Rojas OJ (2010) Cellulose nanocrystals: chemistry, self-assembly, and applications. Chem Rev 110:3479-3500

[14] Sriplai N, Pinitsoontorn S, Bacterial cellulose-based magnetic nanocomposites: A review, Carbohydrate Polymers, Volume 254, 2021,117228,ISSN 0144-8617,https://doi.org/10.1016/j.carbpol.2020.117228.

[15] Blanco Parte G, Santoso S, Chou C, Verma V, Wang HT, Ismadji S, Cheng KC. Current progress on the production, modification, and applications of bacterial cellulose. Critical Reviews in Biotechnology 2020, 40 (3) , 397-414.

[16] Nguyen, N.K.; Nguyen, P.B.; Nguyen, H.T.; Le, P.H. Screening the optimal ratio of symbiosis between isolated yeast and acetic acid bacteria strain from traditional kombucha for high-level production of glucuronic acid. Lwt-Food Sci. Technol. 2015, 64, 1149– 1155

[17] May A, Narayanan S, Alcock J, Varsani A, Maley C, Aktipis A. Kombucha: a novel model system for cooperation and conflict in a complex multi-species microbial ecosystem. PeerJ. 2019 Sep 3;7:e7565. doi: 10.7717/peerj.7565. PMID: 31534844; PMCID: PMC6730531.

[18] Kołodziejczyk A, Kamiński K, Jarosz M, Zastawnik F, Pandyra P, Harasymczuk AG. Bacterial cellulose for clothes production in space using kombucha microbial consortium. 69th International Astronautical Congress (IAC), Bremen, Germany, 1-5 October 2018. Copyright ©2018 by the International Astronautical Federation (IAF). All rights reserved. IAC-18,A1,7,15,x45657

[19] Jayabalan R, Malbasˇa RV, Loncˇar ES, Vitas JS, Sathishkumar M (2014) A review on kombucha tea: microbiology, composition, fermentation, beneficial effects, toxicity, and tea fungus. Compr Rev Food Sci Food Saf 13:538-550

[20] Chakravorty S, Bhattacharya S, Chatzinotas A, Chakraborty W, Bhattacharya D, Gachhui R. Kombucha tea fermentation: Microbial and biochemical dynamics. Int J Food Microbiol. 2016 Mar 2;220:63-72. doi: 10.1016/j.ijfoodmicro.2015.12.015. Epub 2016 Jan 7. PMID: 26796581.

[21] Coton M, Pawtowski A, Taminiau B, Burgaud G, Deniel F, Coulloumme-Labarthe L, Fall A, Daube G, Coton E. Unraveling microbial ecology of industrial-scale Kombucha fermentations by metabarcoding and culture-based methods. FEMS Microbiol Ecol. 2017 May 1;93(5). doi: 10.1093/femsec/fix048. PMID: 28430940.

[22] Lavefve L, Marasini D, Carbonero F. Microbial Ecology of Fermented Vegetables and Non-Alcoholic Drinks and Current Knowledge on Their Impact on Human Health. Adv Food Nutr Res. 2019;87:147-185. doi: 10.1016/bs.afnr.2018.09.001. Epub 2018 Dec 7. PMID: 30678814.

[23] Arıkan M, Mitchell AL, Finn RD, Gürel F. Microbial composition of Kombucha determined using amplicon sequencing and shotgun metagenomics. J Food Sci. 2020 Feb;85(2):455-464. doi: 10.1111/1750-3841.14992. Epub 2020 Jan 20. PMID: 31957879; PMCID: PMC7027524.

[24] Villarreal-Soto SA, Bouajila J, Pace M, Leech J, Cotter PD, Souchard JP, Taillandier P, Beaufort S. Metabolome-microbiome signatures in the fermented beverage, Kombucha. Int J Food Microbiol. 2020 Nov 16;333:108778. doi: 10.1016/j.ijfoodmicro.2020.108778. Epub 2020 Jul 9. PMID: 32731153.

[25] Góes-Neto A, Kukharenko O, Orlovska I, Podolich O, Imchen M, Kumavath R, Kato RB, de Carvalho DS, Tiwari S, Brenig B, Azevedo V, Reva O, de Vera JP, Kozyrovska N, Barh D. Shotgun metagenomic analysis of kombucha mutualistic community exposed to Mars-like environment outside the International Space Station. Environ Microbiol. 2021 Jan 21. doi: 10.1111/1462-2920.15405. Epub ahead of print. PMID: 33476085.

[26] Williams WS, Cannon RE. 1989. Alternative environmental roles for cellulose produced by Acetobacter xylinum. Applied and Environmental Microbiology 55:2448-2452

[27] Smith, C.J., Havenith, G. Body mapping of sweating patterns in male athletes in mild exercise-induced hyperthermia. *Eur J Appl Physiol* **111,** 1391-1404 (2011). https://doi.org/10.1007/s00421-010-1744-8

[28] Martínez-Leal J, Ponce-García N, Escalante-Aburto A. Recent Evidence of the Beneficial Effects Associated with Glucuronic Acid Contained in Kombucha Beverages. Curr Nutr Rep.

2020 Sep;9(3):163-170. doi: 10.1007/s13668-020-00312-6. PMID: 32415557.

[29] Villarreal-Soto SA, Bouajila J, Pace M, Leech J, Cotter PD, Souchard JP, Taillandier P, Beaufort S. Metabolome-microbiome signatures in the fermented beverage, Kombucha. Int J Food Microbiol. 2020 Nov 16;333:108778. doi: 10.1016/j.ijfoodmicro.2020.108778. Epub 2020 Jul 9. PMID: 32731153.

[30] Mousavi SM, Hashemi SA, Zarei M, Gholami A, Lai CW, Chiang WH, Omidifar N, Bahrani S, Mazraedoost S. Recent Progress in Chemical Composition, Production, and Pharmaceutical Effects of Kombucha Beverage: A Complementary and Alternative Medicine. Evid Based Complement Alternat Med. 2020 Nov 18;2020:4397543. doi: 10.1155/2020/4397543. PMID: 33281911; PMCID: PMC7688354.

[31] Blanc PJ (1996) Characterization of the tea fungus metabolites. Biotechnol Lett 18:139-142

[32] Marsh AJ, O'Sullivan O, Hill C, Ross RP, Cotter PD (2014) Sequence-based analysis of the bacterial and fungal compositions of multiple kombucha (tea fungus) samples. Food Microbiol 38:171-178

[33] Kamiński, K., Jarosz, M., Grudzień, J. *et al.* Hydrogel bacterial cellulose: a path to improved materials for new eco-friendly textiles. *Cellulose* **27**, 5353-5365 (2020). https://doi.org/10.1007/s10570-020-03128-3

[34] Arau´jo S, da Silva FM, Gouveia IC (2015) The role of technology towards a new bacterial-cellulose-based material for fashion design. J Ind Intell Inf 3:168-172

[35] Salmeia KA, Jovic M, Ragaisiene A, Rukuiziene Z, Milasius R, Mikucioniene D, Gaan S (2016) Flammability of cellulose based fibers and the effect of structure of phosphorus compounds on their flame retardancy. Polymers 8:293(1)– 293(15)

[36] Wang HY, Qian H, Yao WR (2011) Melanoidins produced by the mail lard reaction: structure and biological activity. Food Chem 128:573-584

[37] Kulkarni PK, Anil Dixit S, Singh UB (2012) Evaluation of bacterial cellulose produced from Acetobacter xylinum as pharmaceutical excipient. Am J Drug Discov Dev 2:72-86

[38] Bandyopadhyay S, Saha N, Saha P (2018) Characterization of bacterial cellulose produced using media containing waste apple juice. Appl Biochem Biotechnol 54:649-657

[39] Akkus A, Tyler R, Schiraldi D, Roperto R, Faddaul F, Teich S (2017) Effect of polyethylene oxide on the thermal degradation of cellulose biofilm: low cost material for soft tissue repair in dentistry. J Clin Exp Dent 9:e875–e878

[40] Andrade FK, Silva JP, Carvalho M, Castanheira EM, Soares R, Gama M (2011) Studies on the hemocompatibility of bacterial cellulose. J Biomed Mater Res A 98:554-566

[41] Swingler S, Gupta A, Gibson H, Kowalczuk M, Heaselgrave W, Radecka I. Recent Advances and Applications of Bacterial Cellulose in Biomedicine. Polymers (Basel). 2021 Jan 28;13(3):412. doi: 10.3390/polym13030412. PMID: 33525406; PMCID: PMC7865233.

[42] Andriani, D., Apriyana, A.Y. & Karina, M. The optimization of bacterial cellulose production and its applications: a review. *Cellulose* **27**, 6747-6766 (2020) https://doi.org/10.1007/s10570-020-03273-9

[43] Mustafa F, Andreescu S. 2020. Nanotechnology-based approaches for

food sensing and packaging applications. DOI: 10.1039/D0RA01084G (Review Article) RSC Adv., 2020, **10**, 19309-19336

[44] Global microbial and bacterial cellulose market research report, 2020

[45] Azeredo HMC, Barud H, Farinas CS, Vasconcellos VM, Claro AM (2019) Bacterial cellulose as a raw material for food and food packaging applications. Front Sustain Food Syst 3:1-14

[46] Costa AFS, Almeida FCG, Vinhas GM, Sarubbo LA (2017) Production of bacterial cellulose by Gluconoacetobacter hansenii using corn steep liquor as nutrient sources. Front Microbiol 8:1-12

[47] Kozyrovska N, Reva O, Podolich O. Et al., 2021. To other planets with upgraded millennial kombucha in rhythms of sustainability and health support. In print.

[48] https://www.nasa.gov/vision/space/livinginspace/ Astronaut_Laundry.html

[49] Podolich O, Zaets I, Kukharenko O et al (2017) Kombucha multimicrobial community under simulated spaceflight and martian conditions. Astrobiology 17:459-469

[50] Podolich O, Kukharenko O, Haidak A et al (2019) Multimicrobial kombucha culture tolerates Mars-like conditions simulated on low Earth orbit. Astrobiology 19:183-196

Section 2

Current Scenario

Chapter 6

Biorefinery System of Lignocellulosic Biomass Using Steam Explosion

Chikako Asada, Sholahuddin and Yoshitoshi Nakamura

Abstract

Recently, plant biomass has been attracting attention due to global warming and the depletion of fossil fuels. Lignocellulosic biomass (i.e., wood, straw, and bagasse) is attracting attention as an abundant renewable resource that does not compete with the food resources. It is composed of cellulose, hemicellulose, and lignin and is a potential resource that can be converted into high-value-added substances, such as biofuels, raw materials for chemical products, and cellulose nanofibers. However, due to its complicated structure, an appropriate pretreatment method is required for developing its biorefinery process. Steam explosion is one of the simplest and environmentally friendly pretreatments to decompose lignin structure, which converts cellulose into low-molecular-weight lignin with high efficiency. It has received significant attention in the field of not only biofuel but also biochemical production. Steam explosion involves the hydrolysis of plant biomass under high-pressure steam and the sudden release of steam pressure induces a shear force on the plant biomass. Moreover, it is a green technology that does not use any chemicals. Thus, a steam explosion-based biorefinery system is highly effective for the utilization of lignocellulosic into useful materials, such as ethanol, methane gas, antioxidant material, epoxy resin, and cellulose nanofiber.

Keywords: steam explosion, biorefinery, lignocellulosic, environmentally friendly, conversion

1. Introduction

The agricultural and forestry sectors produce the maximum amount of lignocellulosic biomass waste with minimum utilization. Moreover, many abundant crops, such as bamboo, weed, and shrubs, are not utilized optimally. Recently, lignocellulosic biomass has been widely used for burning fuel, such as gasification through the palletization process; however, the burning pathway in this process could only result in heat and a minimum amount of byproduct. Thus, the process of obtaining byproducts from lignocellulosic biomass requires considerable improvement. The utilization pathway of renewable sources for biofuels has been improved through the second generation of biofuels obtained from nonfood resources. However, the biofuel production process still results in potential waste that could have been used for obtaining another product. This opens up possibilities for the total conversion of biomass into biomaterials, pharmaceuticals, and raw chemicals with minimum waste.

Lignocellulosic biomass contains three main sources of core materials that can be developed into more valuable products: lignin, cellulose, and hemicellulose. Lignin performs biological functions such as providing mechanical support for resistance to various stresses and water transport. It is a highly irregular polymer of phenol subunits. Phenol is a derivative of the organic molecule benzene that exhibits high stability with the alcohol functional group. Lignin has recently become a promising raw material that can be converted into various products, such as biopolymer-based products (i.e., lignin-epoxy resin (LER), polyphenols sources, antioxidants, and various chemicals (i.e., 5-hydroxymethyl furfural (HMF) and furfural). Meanwhile, cellulose is a promising raw material for second-generation biofuel and has been widely used as a source of biomaterials, such as cellulose nanofiber (CNF), which can change the dominancy of conventional fillers for composites and performs other functions in biomedical and other industries. Hemicellulose has long been used as a source for monosaccharides, such as xylose, which can produce xylitol, and various biochemicals.

Lignocellulosic biomass is difficult to treat, wherein the dominance of lignin carbohydrate-complex (LCC) causes the recalcitrance of the separation process. Steam explosion is an environmentally friendly physiochemical pretreatment widely used for lignocellulosic biomass, which only uses water. This section illustrates the biorefinery process of obtaining lignocellulosic biomass from various sources and provides information on developments made in the use of steam explosion pretreatment to convert lignocellulosic biomass into value-added products, such as various biofuels, derivative methane gas, biopolymers, raw material, biochemical, and polyphenol.

2. Steam explosion pretreatment

In 1926, the steam explosion was introduced to efficiently process lignocellulosic biomass into fibers [1, 2], and in 1932, Babcock was patented as the first method for producing sugar and alcohol from wood chips under high pressure and temperature with varying retention time, where the chips were suddenly discharged through

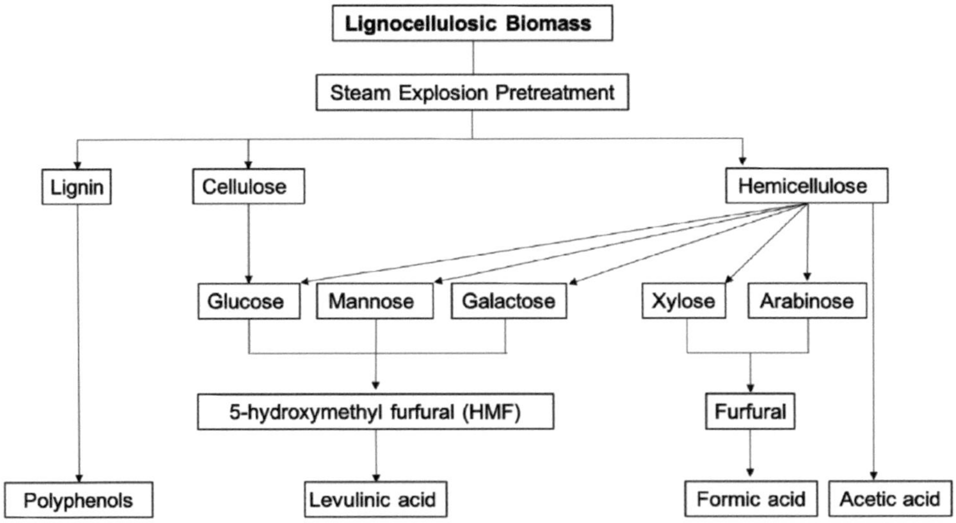

Figure 1.
Route changes in the steam-exploded lignocellulosic compound.

a slotted port and an explosion was generated [3]. Steam explosion pretreatment is still being extensively used and increasing the scope of lignocellulosic biomass utilization.

Steam explosion is classified as a psychochemical pretreatment that can change a compound and chemical structure. The autohydrolysis mechanism involved in steam explosion facilitates mechanical cutting and fiber defibrillation, which hydrolyze some of the cell-wall components and release acid in the process. The structural components are broken by the diffusion process under the induced pressure and heat that penetrate the recalcitrant of the cell wall; furthermore, the pressure released in the explosion process separates the fibers through sudden evaporation of the condensed moisture. As shown in **Figure 1**, during the steam explosion pretreatment, the biomass changes the main route as follows: lignin to low-molecular-weight lignin (LML) and polyphenols; cellulose to glucose; and hemicellulose to acetic acid, glucose, xylose, and other monosaccharides. Glucose, mannose, and galactose continue to be converted into HMF, and levulinic acid, xylose, and arabinose continue to be converted into furfural and formic acid.

3. Conversion concept

Various conversion concepts using steam explosion have been introduced to obtain lignocellulosic byproducts, such as raw materials and biochemicals, as described in **Figure 2**. Generally, lignin is converted into LML and polyphenols, curing agent, and LER; holocellulose is converted into CNF and biofuels, such as ethanol and biogas; and hemicellulose is converted into monosaccharides and their derived products.

3.1 Green conversion

To meet the current requirements of obtaining environmentally friendly byproducts, green conversion has been introduced. This has emphasized the need of a biorefinery method that can reduce the amount of waste generated by using the sustainable development goal (SDG) program, which reduces the environmental

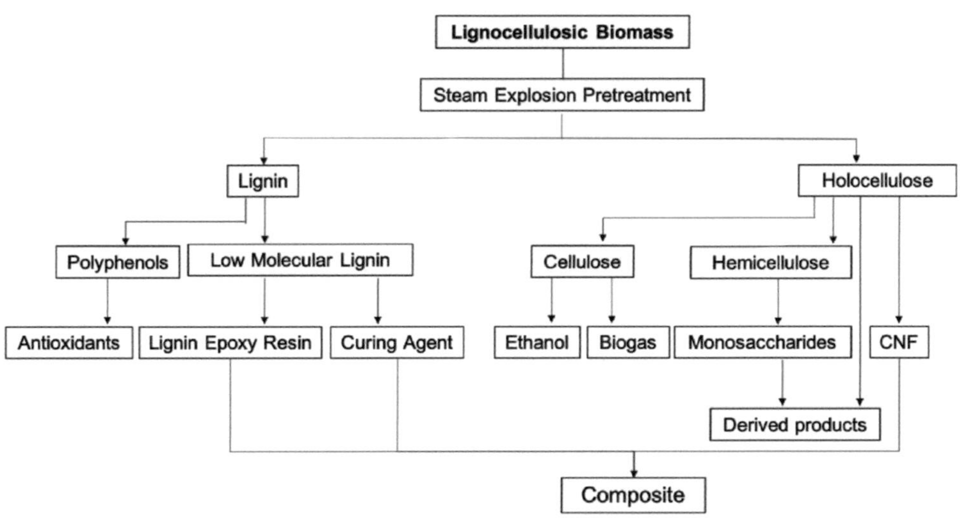

Figure 2.
Conversion scheme for the main routes of steam-exploded lignocellulose.

impact of global warming. Steam explosion pretreatment is considered environmentally friendly because it only uses pressure and water. To obtain a green conversion, various green biorefinery processes have been proposed, which produce minimum waste and reusable chemicals during the process. For example, for the conversion of lignocellulosic biomass into raw materials and biofuels, green conversion using steam explosion has been extensively reported. Ethanol production from lignocellulosic biomass follows a basic pathway: steam explosion–enzymatic saccharification–ethanol fermentation under varying steaming times and pressures with various additional methods, such as the combination of steam explosion with other pretreatments; reduction of the fermentation inhibitor, various enzymes, and their dosages; and the hydrolysis and fermentation process. Another biorefinery process for obtaining cellulose and hemicellulose byproducts, such as monosaccharides, uses the basic methods of enzymatic saccharification and combination pretreatment for ethanol production.

3.2 Total biorefinery

The total biorefinery concept for lignocellulosic biomass has been introduced to maximize the amount of byproduct produced from each biorefinery process and reduce waste production. A conventional biorefinery focuses on only one product from lignocellulosic biomass, such as derived products of cellulose, hemicellulose, or lignin. The waste generated from a conventional biorefinery contains a potential raw material, which is wasted into effluents; for example, the waste generated from lignocellulosic biomass ethanol production still contains lignin, which could have been used as lignin-derived products.

Asada [4] introduced a waste reduction system to obtain more useful products through steam explosion pretreatment, followed by water and methanol extraction. They used a water-soluble material for the purification process to obtain monosaccharides and oligosaccharides and methanol-soluble lignin for the resinification process to obtain LER. The two solid residues (i.e., hollocellulose and klason lignin) were used to obtain antibacterial violet pigment or lactic acid and activated carbon, respectively. The antibacterial violet pigment was produced using the enzyme saccharification process to obtain the monosaccharide content, followed by lactic acid fermentation using *Lactobacillus plantarum* and *Janthinobacterium lividum*. The waste generated from enzymatic saccharification (i.e., klason lignin) was processed into activated carbon through carbonization in a furnace at 500°C under a nitrogen gas atmosphere. Hongzang [5] examined the steam explosion-based total biorefinery process, followed by washing with water and alcohol extraction. They used a water-soluble material for the fermentation process and purified an alcohol-soluble material to obtain LML. The solid residue generated from the alcohol extraction was subjected to the pulping process, with cellulose as the final product. Asada [6] subjected Japanese cedar (*Cryptomeria japonica*) to steam explosion pretreatment followed by water extraction and methanol extraction. They used all residues obtained from each process to obtain the potential products; for example, they used the water-soluble material from water extraction for obtaining antioxidant resources by examining the antioxidant activities. Furthermore, the methanol-soluble lignin obtained from methanol extraction produced LER through epoxy resin synthesis and the hollocellulose obtained from the solid residue of methanol extraction was used for ethanol production using simultaneous saccharification and fermentation process (SSF). In another study, Asada [7] subjected cedar (*Cryptomeria japonica*), eucalyptus (*Eucalyptus globulus*), and bamboo (*Phyllostachys pubescens*) to steam explosion pretreatment, which produced raw materials by the continuous biorefinery process. They used water extraction, which produced a water-soluble material rich in polyphenol content; the residue obtained from this process was used to continue the methanol extraction with LML as the raw material, and finally, enzymatic saccharification with glucose as the raw material was performed using the residue obtained from methanol extraction. LML was subjected to epoxy resin synthesis to produce LER. In another study, Asada [8] reported the steam explosion-based total biorefinery process for lignocellulosic biomass, followed by water extraction and acetone extraction, to produce a phenolic compound as an antioxidant, followed by acetone extraction, which produced LML. This LML was converted into LER and a curing agent. The residue obtained from the acetone extraction was used to convert hollocellulose (cellulose and hemicellulose) into CNF, with the cured epoxy resin as an end product. The final product of LER, curing agent, CNF was producing the cured epoxy resin.

4. Biofuels

Biofuels face competition with other food resources such as wheat, grain, corn, cassava, and palm oil. To address this issue, researchers started evaluating other renewable resources that do not face this competition. The second generation of biofuels is obtained from unutilized biomass and biomass waste resources, which are abundantly available. The biomass waste generated from agricultural and forestry sectors, such as burning fuel or dumping of disposal waste, commonly ends up with minimum utilization. Moreover, the unutilized biomass, such as weed, grass, and bamboos, can be converted into biofuels. The second-generation biofuels have been widely introduced with various biorefinery processes, including steam explosion pretreatment for bioethanol and biogas production.

4.1 Ethanol

The second-generation ethanol production is highly dependent on the availability of cellulose and hemicellulose from LCC. Steam explosion is an effective pretreatment for breaking LCC into cellulose and hemicellulose, which can be readily converted into ethanol. This pretreatment is widely used because of its ability to break LCC into cellulose and hemicellulose obtained from the conversion of lignocellulosic biomass into ethanol. The general routes for ethanol production via steam explosion pretreatment are described in **Figure 3**. Nakamura [9] converted rice straw into ethanol through steam explosion, followed by water extraction and enzymatic saccharification and fermentation. They reported 86% theoretical ethanol production from the substrate by using steam explosion and *Trichoderma viride* cellulase or Meicelase and *Aspergillus aculeatus cellulase* or Acucelase, using *Pichia stipites*. Sasaki [10] converted wood chips and acorns of *Quercus acutissima* into acetone, butanol, ethanol (ABE) by steam explosion pretreatment, followed by direct conversion using the separate hydrolysis and fermentation (SHF) and SSF methods. The SSF method yielded 100% ABE conversion rate, with 0.112 g/L/h ABE productivity for 196 h of incubation time with 60 g/l concentration of the initial substrate. In another study, Sasaki [11] subjected the pruned branches of pear trees to steam explosion pretreatment, followed by SSF, to produce ethanol. They compared the steam-exploded substrates with and without water and methanol extraction in terms of the amount of ethanol produced. The result showed that the SSF with water and methanol extraction achieved 76% of the theoretical production with 99.7% of glucose recovery, whereas the sample not subjected to methanol and water extraction majorly produced sugar and no ethanol. This result

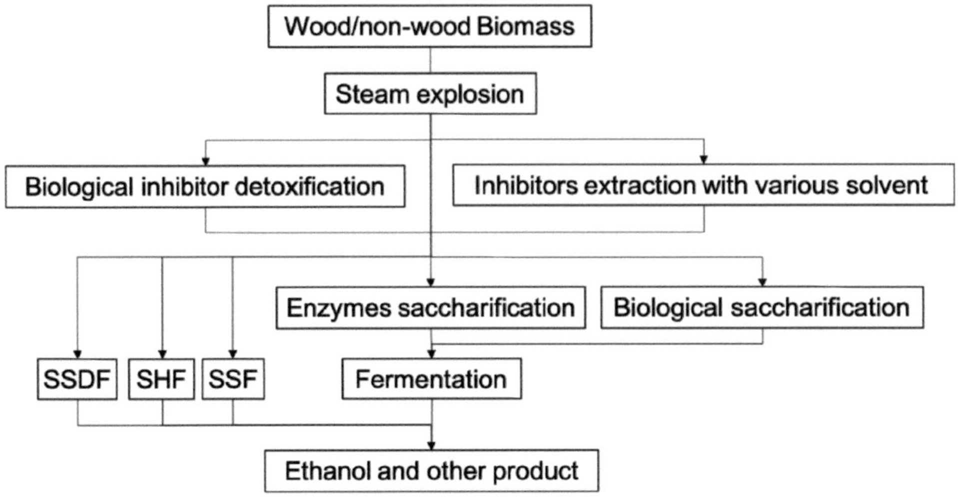

Figure 3.
Production routes for ethanol produced from steam explosion.

was attributed to fermentation inhibitors contained in the substrate as compound fractions obtained from steam explosion pretreatment.

Table 1 shows the steam explosion-based ethanol production methods. The most significant problem incurred in ethanol production using steam explosion pretreatment is the yield of a fermentation inhibitor as a derived product of steam explosion. Ando [24] identified the influences of an aromatic monomer in steam explosion on ethanol production via *Saccharomyces cerevisiae*, which could be reduced by washing the inhibitors with a solvent, converting them into inactive compounds using biological or chemical methods, improving the steam explosion conditions to minimize the inhibitor formation, and screening for yeasts that resist the inhibitors. Asada [19] used the SSF method for spent of shitake mushroom medium for evaluating the effect of reducing the formation of fermentation inhibitors with and without water extraction. They reported 87.6% of theoretical ethanol yield, which produced 15.9 g ethanol from 100 g substrate. In another study, Asada [15] used steam explosion, followed by water extraction, in comparison with mechanical grinding with a ball mill, for converting disposable chopstick obtained from aspen into ethanol. This resulted in 20 FPU/g samples, and the continuous fermentation using the SSF method resulted in 520 and 598 mg-glucose/g-dry samples from grinding and steam explosion pretreatment, respectively. This method yielded 79% of theoretical ethanol production with 241 mg-ethanol/g-dry. In addition, Scholl [20] used steam explosion pretreatment for converting the elephant grass *Pennisetum purpureum* under various pressures and steaming times, followed by vacuum draining and washing with water to remove the inhibitor material and enzymatic saccharification using cellulase and xylanase obtained from *P. echinulatum*. Thereafter, the fermentation process was continued using *Saccharomyces cerevisiae* as the steam-exploded elephant grass, followed by water washing and using cellulose enzymes with 10 FPU/g total solid having 4 wt% substrate concentration, which did not result in a feasible production (110.45 μl/g). To overcome the toxicity of phenolic compounds obtained from the steam-exploded substrate, which act as an inhibitor for ethanol production, various treatments have been reported, such as the use of laccase produced from *Coriolopsis rigida* and *Trametes villosa*. The detoxification by laccases considerably improves the ethanol fermentation, which in turn reduces the toxic effect on *S. cerevisiae* [18]. Asada [17] converted cedar into ethanol by using steam explosion pretreatment in

DOI: http://dx.doi.org/10.5772/intechopen.98544

Biomass	Pretreatment	Pressures (MPa)	Steaming time (min)	Inhibitors removal	Method*	Enzyme and/or microorganism			Achievement (Glucose or ethanol yields)	References
						Saccharification	Fermentation	Other		
Potato	Steam explosion	0.51 to 3.04	10	n/a	BS, SSF	*A. Awamori*	*S. cerevisiae*		87%	[12]
Rice straw	Steam explosion	3.0 to 4.0	0.5 to 10	n/a	ES, SHF	Acucelase, Meicelse	*Pichia stipitis*		86%	[9]
Domestic waste	Steam explosion	0.5 to 2.5	3	n/a	ES, SSF	*A. awamori*, *Trichoderma viride* NBRC 31327	*S. cerevisiae*		n/a	[13]
Japanese cedar	Steam explosion, organosolv	2.5 to 4.5	5	n/a	ES, SSF	Meicelse	n/a		84.4%	[14]
Aspen chopsticks	Steam explosion	2.0 to 3.0	5	Water extraction	ES, SSF	Meicelse				[15]
Sawdust waste	Steam explosion	2.5 to 5.5	5	Water & Methanol Extraction	ES, SSF	Meicelase, *Trichoderma viride*, β-glucosidase	*S. cerevisiae*		81%	[16]
Japanese cedar	Steam explosion	3.5 to 5.5	5 & 10	Water extraction, Biological detoxification	ES, SSF, SSDF	Meicelase	*S. cerevisiae* BA11	*U. thermosphaericus* A1	88%, 74%	[17]
Q. acutissima	Steam explosion	3.5	5	Water & Methanol Extraction	ES, SSF, SHF	Glutase-AN, Cellic CTec 2	*C. acetobutylicum* NBRC 13948		100%	[10]
Pear trees	Steam explosion	2.5 to 3.5	5	Water & Methanol Extraction	ES SSF	Cellic CTec 2	*Kluveromyces marxianus* NBRC 1777		76%	[11]

Biomass	Pretreatment	Pressures (MPa)	Steaming time (min)	Inhibitors removal	Method*	Enzyme and/or microorganism			Achievement (Glucose or ethanol yields)	References
						Saccharification	Fermentation	Other		
Japanese cedar	Steam explosion	4.5	3	Water & Methanol Extraction	ES, SSF	Meicelase, *Trichoderma viride*, β-glucosidase	*S. cerevisiae*		70%	[6]
Wheat straw	Steam explosion	n/a	n/a	n/a	ES, Fermentation	Laccase	n/a	*C. rigida* & *T. villosa*	n/a	[18]
Sawtooth Oak	Steam explosion	1.0 to 3.0	5	Water extraction	SSF	Meicelase	*S. cerevisiae* AM12		87.6%	[19]
Elephant grass	Steam explosion	1.27 to 2.05	6 to 10	Vacuum dring, Water washing	ES, Fermentation	*P. echinulatum* S1M29	*S. cerevisiae CAT-1*		n/a	[20]
Lemon peel	Steam explosion	0.6	5	n/a	SSF, SHF	Pectinase, Cellulase, β-glucosidase (Novozym 188)	*S. cerevisiae* CECT 1329		n/a	[21]
Brassica napus	Steam explosion, H2SO4, CaO	n/a	n/a	n/a	ES, Fermentation	Cellulases, Xylanase	*S. cerevisiae*		90–93%	[22]
Bagasse	Steam explosion	1.57 to 3.53	1 to 10	Water extraction	ES, Fermentation	Meicelase, Acucelase	*Pichia stipitis* CBS 5773		n/a	[23]

Table 1.
Steam explosion-based ethanol production methods.

Footnote to Table 1: **BS: Biological saccharification, ES: Enzymatic saccharification, SHF: Separate hydrolysis and fermentation, SSF: Simultaneous saccharification and fermentation, SSDF: Simultaneous saccharification, detoxification, and fermentation.*

comparison with water and methanol extract, followed by the use of the simultaneous saccharification, detoxification, and fermentation (SSDF) method combined with the detoxification process using *U. thermosphaericus* A1 as a biological fermentation inhibitor degrading agent to degrade the inhibitory material produced from the steam explosion pretreatment. They concluded that the fermentation inhibitor degradation in low concentration was necessary to produce ethanol from steam-exploded lignocellulosic biomass, where the steam-exploded cedar produced glucose alone due to the saccharification end product.

4.2 Biogas

Figure 4 shows the cellulose conversion into methane gas. The methanogenic reaction of cellulose or other forms of carbohydrates through the hydrolysis process yields monosaccharides, which are fermented to produce acetic acid, hydrogen, and CO_2 [25]. From the fermented products, acetolactic methanogenesis converts the acetic acid and hydrogenotrophic methanogenesis converts the hydrogen and CO_2; the conversion process includes reforming the acetic acid, hydrogen, and CO_2 into methane [26]. Steam explosion pretreatment has been widely used for biogas production. Take [27] subjected Japanese cedar chips to 4.51 MPa steam explosion for 5 min for optimal methane production. Kobayashi [28] used bamboo to produce methane by 3.53 MPa steam explosion (243°C) using the sludge obtained from sewage treatment for microbial seed under mesophilic condition (37°C), which resulted in 80% theoretical yield with 423 ml obtained from 1 gr of cellulose and hemicellulose for 25 days of total cumulative production. Mulat [29] combined steam explosion pretreatment for lignocellulosic biomass and bioaugmentation using *Caldicellulosiruptor bescii*, which enhanced the methane production under thermophilic conditions by 140% in 50 days with low dosages of *Caldicellulosiruptor bescii* inoculum (2–5%). Sholahuddin [30] subjected rice husk to a combination of steam explosion pretreatment at 2.52 MPa and 224°C followed by water extraction and activated cow dung as the inoculum without co-digestion, at 37°C. This yielded 96.1% of stochiometric prediction of methane production with 199 ml/g of total solid, which contained 41% of cellulose only for 22 days, and all the liquid and solid residues were used as the substrate. Steam explosion pretreatment was also used for grass, such as reed, which can be used as a potential raw material for biogas because of its abundance. Lizasoain [31] subjected reed biomass (*Phragmites australis*) for biogas feedstock to steam explosion under various temperatures, pressures, and steaming times, where the 200°C and 15 min combination increased the methane yield by 85% compared to the untreated samples. Furthermore, Dererie [32] used oat straw for combined biogas and ethanol production with steam explosion pretreatment and other chemical treatment; as per the result, the residue of ethanol fermentation from steam-exploded oat straw produced higher methane than that produced by unfermented steam-exploded oat straw. They concluded that the fermentation ethanol process acts as an additional pretreatment for methane production.

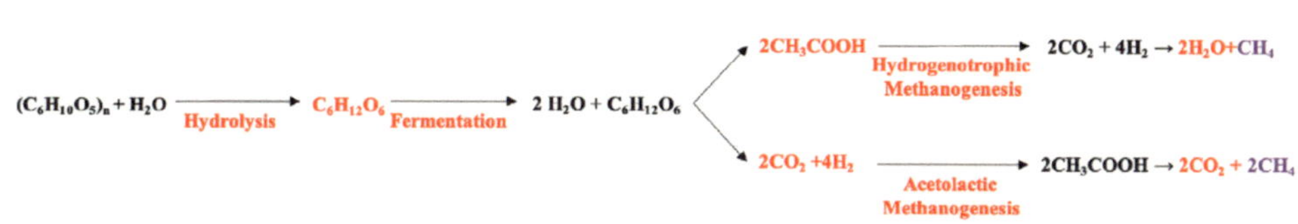

Figure 4.
Cellulose conversion into methane gas.

Methane production from lignocellulosic feedstock through steam explosion pretreatment provides a wide spectrum of total conversion. Methane is converted not only from cellulose and hemicellulose but the aromatic lignin fractions also contribute to the methane production. Moreover, steam explosion facilitates better anaerobic digestion by disrupting the lignin structure [33], which can be converted into methane. However, the anaerobic degradation of the aromatic compound incurs several difficulties in the degradation process [34], and several studies have reportedly observed anaerobic lignin degradation [35, 36]. The aromatic lignin heteropolymers mainly comprise two monolignols, which are methoxylated to various degrees: synapyl and coniferile alcohols; these monolignols are fused into lignin in the unit syringyl (S) and guaiacyl (G) forms, respectively [37]. The depolymerized monomeric unit of lignin (i.e., S) is converted into vanillin and the G unit is converted to syringaldehyde [38]. Syringaldehyde can produce a high methane yield [19, 24, 39]. Barakat [40] demonstrated the combination of xylose and the lignin fraction, such as aromatic compounds syringaldehyde and vanillin; cellulose and hemicellulose fractions, such as HMF; and furfural and xylose. The combination of xylose and syringaldehyde yielded the highest methane production, followed by the combination of xylose and furfural compared with xylose alone.

5. Cellulose- and hemicellulose-derived products

The cellulose- and hemicellulose-derived products by steam explosion pretreatment, such as monosaccharides, antibacterial, lactic acid, D-lactic acid, xylose, xylitol, HMF, and furfural, have been widely introduced in many studies. The production of monosaccharides as a raw material derived from cellulose- and hemicellulos-based products, such as glucose conversion, has been reported. Fernandez [41] reported the use of steam explosion for whole olive stones to enhance the enzymatic hydrolysis for conversion of cellulose into sugar. Dai [42] used steam explosion, in comparison with hydrothermal pretreatment, to convert bamboo (*Bambusa stenostachya*) into monosaccharides, such as glucose xylose and arabinose and furfural. Asada [4] reported the production of lactic acid and antibacterial through enzymatic saccharification followed by the fermentation process, and as a result, approximately 0.53 g/l where the lactic acid from glucose produced approximately 0.7 g/l from 1 g/l glucose. This result was attributed to the inhibition of the fermentation process of soluble lignin from the substrate of enzymatic hydrolysate. The antibacterial violet pigment was produced by the same process of enzymatic hydrolysate, where glucose decreased in line with the rapid cell growth, with the maximum growth occurring within 20 h. After the cell growth was stopped, the antibacterial was produced as a secondary metabolite and reached the maximum condition within 30 h, where the culture color changed into deep violet. Sasaki [43] used steam-exploded sugarcane bagasse to produce D-lactic acid by comparison with and without water extraction. They used meicelase for enzymatic saccharification and *Lactobacillus delbrueckii* NBRC 3534 for D-lactic acid fermentation; according to the result, the glucose conversion rate for water-extracted hydrolysate (49.9%) was found to be higher than that obtained without extraction (36.4%). In terms of the rate of conversion from glucose to D-lactic acid, D-lactic acid production was escalated more by the water extraction treatment (90%) than that without the extraction (66.6%). This result was attributed to the water extraction, which was decreased by fermentation inhibitors such as HMF and furfural. Macros [44] used wheat straw to produced xylose and glucose by the steam explosion pretreatment followed by warm water wash and enzymatic hydrolysis, which used a combination of accellerase and xylanase (accellerase-XC and XY), which exhibits multiple enzyme activities, such as exoglucanase, endoglucanase, hemicellulose, and β-glucosidase. As a result, 87% and 27% increases were observed in the hemicellulose conversion into glucose and xylose, respectively. Liu [45] used simultaneous and co-fermentation for improving the xylose production by steam explosion, followed by SHF and SSCF methods. *S cerevisiae* IPE002 was used for both fermentation methods, which could convert. The glucose and xylose production reached 100.4 and 25.1 g/l concentrations, respectively. Wang [46] used corn straw to produce xylitol by acid impregnation, followed by steam explosion without detoxification; as a result, xylene hydrolysate with a high amount of xylose production and simultaneous glucan production were observed. The medium of *Candida tropicalis* produced a maximum yield of 35.6 g/l for xylitol in 38 h fermentation with 0.94 g/l h productivity.

6. Biomaterials

The lignocellulosic biomaterial has the potential to be used as a raw material for developing a new renewable and environmentally friendly product. The steam explosion pretreatment is a highly recommended pretreatment to obtain more valuable raw materials in the biorefinery process due to its effective breaking process, which yields lignin, cellulose, and hemicellulose byproducts. The lignocellulosic biomaterial can be used in various fields, such automobiles, medical, pharmaceutical, food packaging, beverage cans, electronics, composite industries, and the aerospace industry. The lignocellulosic biomaterials such as LER as biopolymers exhibit valuable properties, such as low moister absorption, good mechanical and electrical properties, and high chemical and thermal resistance. CNF is widely known as a raw material that exhibits good optical and mechanical properties, low thermal expansion coefficient, and high specific area. The basic properties of the raw material of lignocellulosic make it a promising raw material.

6.1 CNF

CNF is a promising biomaterial material that has advantageous characteristics beside those of optically transparent functional material [47], same as plastic (i.e., high gas barrier properties [48], biodegradability, light weight, high strength, ultrafine fibers, large specific surface area, low thermal expansion, characteristic viscosity in water, and environmentally friendly biomaterials, which can be used for cosmetic, biomedical, and pharmaceutical products [49–51], nanocomposites for industrial products [52], and filters that have large surface area for collecting small dust particles [53]). **Table 2** lists an extensively reported substrate developer for CNF resources; wood and nonwood biomass is the most potential CNF resource owing to the considerable abundant waste generated from the wood industry and agricultural waste.

Before isolating CNF, it is necessary to break the recalcitrant of the compact-structure LCC from wood and nonwood lignocellulose biomass before it is used as a CNF source. There are several treatments to break the LCC; in this study, steam explosion pretreatment is recognized as a promising method to obtain CNF due to its effectiveness in the biorefinery process and being considered as environmentally friendly.

After wood or non-wood lignocellulosic biomass is treated with steam explosion, two main routes have been proposed, as described in **Figure 5**. The first route is the use of the extraction process to separate the other compounds that could be used as other biomaterials, such as LML and polyphenols, by water, methanol, or acetone extraction as part of the delignification process. Then, the lignin still attached to hollocellulose is separated by the bleaching process. Several bleaching

Biomass	Pretreatment	Condition		Delignification	Nanosizer	References
		Pressure (MPa)	Steaming time (min)			
Abaca (*Musa textilis*)	Steam explosion	0.13	15	Water, Homogenizer	Oxidation	[54]
Japanese cedar	Steam explosion	3.0 to 4.0	5 to 15	Water, Acetone, Bleaching (NaClO2), NaOH	Grinder	[55]
Yerba mate	Steam explosion	0.15	60	Water, Hydrogen peroxide, Water	Oxalic acid	[56]
Sugarcane bagasse	Steam explosion	1.3 to 1.9	5 to 15	Water, Immersed hydrogen peroxide, NaOH	High-speed agitation, High pressure homogenization	[57]
Wheat straw	Steam explosion	3	2	Water, Microwave alkali hydrolysis, NaOH	Microfluidization	[58]
Corncorbs	Alkaline steam explosion	n/a	n/a	NaOH, Bleaching (NaClO2)	High-speed agitation, Ultrasonication	[59]
Pineapple leaf	Alkaline steam explosion	0.13	60	NaOH, Acetic acid, Bleaching (NaClO2), Oxalic acid, KMnO4	High speed mechanical stirring	[60]
Sugarcane bagasse	Alkaline steam explosion	0.1	300	NaOH, Hydrogen peroxide, Water, HCl	Ultrasonication	[61]
Sugar beet pulp	Steam explosion	2.4	35	Hydrogen peroxide, Water	High-speed agitation, Ultrasonication	[62]
Banana fibers	Alkaline steam explosion	0.13	60	NaOH, Acetic acid, Bleaching (NaClO2), Oxalic acid, KMnO4	High speed mechanical stirring	[63]
Abaca (*Musa textilis*)	Steam explosion	0.13	15	HNO3, NaNO2, NaSO3, Water, Hydrogen peroxide	Hydrolyzed H2C2O4, Homogenization	[64]
Palm Oil Empty Fruit Bunches	Steam explosion	2	10	Sodium hydroxide, Water, Bleaching (NaClO2)	Grinder	[65]
Wheat straw	alkaline steam explosion	0.1	240	NaOH, Immersed hydrogen peroxide, HCl, Ultrasonicator, Water	Homogenizer	[66]

Biomass	Pretreatment	Condition		Delignification	Nanosizer	References
		Pressure (MPa)	Steaming time (min)			
Sugarcane bagasse	Steam explosion	1.1 to 1.5	n/a	Catalyzed hydrothermal NaOH, Bleaching, Hydrogen peroxide, Water	High-speed agitation, Ultrasonication	[67]
White poplar	Steam explosion	2.5 to 3.5	1 to 5	Bleaching (NaClO2), Acetic acid	Grinder	[8]
Rice straw	Steam explosion	n/a	n/a	NaOH, Immersed hydrogen peroxide	TEMPO-oxidized	[68]

Table 2.
CNF production via steam explosion pretreatment.

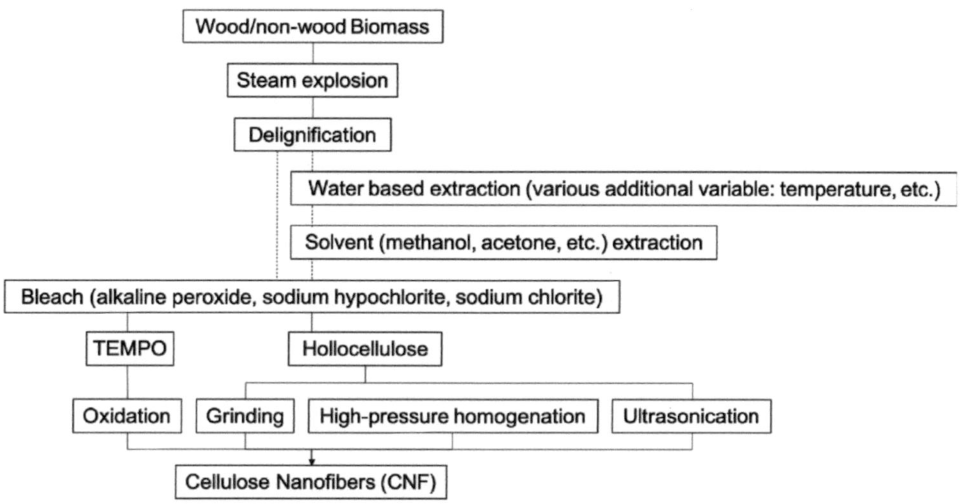

Figure 5.
CNFs produced from steam explosion-based general route biorefinery.

agents are available in the market, such as alkaline peroxide, sodium hypochlorite, and sodium chlorite, which are separated by filtration to obtain holocellulose, which is a raw material used for CNF. Hollocellulose can be directly processed to obtain CNF by using the grinding treatment [55, 69–71], high-pressure homogenization [57, 72–74], acid hydrolysis [54, 56], enzymatic hydrolysis [41, 75], and ultrasonication [67, 76, 77]. The second route is the bleaching process followed by TEMPO-mediated oxidation to isolate the CNF [68].

6.2 Biopolymers

The production of biopolymers through steam explosion pretreatment is a potential and feasible biorefinery process that can produce various polymer raw materials, such as a curing agent and lignin resin, and byproducts, such as LER and composite material [78]. A biodegradable polymer obtained from renewable resources has recently attracted attention as a substitute for hegemony petroleum-based polymers, which support the SDG program to reduce the effect of global warming. LER production from steam-exploded lignocellulosic biomass has been extensively reported. Lignocellulosic biomass is a potential resource for epoxy resins derived from lignin and hemicellulose (i.e., lignin-based epoxies, furfural-based epoxies, and phenolic and polyphenolic epoxies). Lignin-based epoxies include depolymerized organosolv lignin [79], depolymerized hydrolysis lignin [80], LER, which acts as both a curing agent and a resin [4, 6, 7, 81–84], diglycidyl ether of vanillyl alcohol/IPDA, diglycidyl ether of methoxyhydroquinone, diglycidyl ether of vanillic acid [85], and vanillin-based epoxies [86]. Furfural-based epoxies include furan diepoxide and bis-furan di-epoxide furan monoepoxide, 2,5-Bis[(2-oxiranylmethoxy)methyl]-furan [87], 1,4-Bis[(2-oxiranylmethoxy)methyl]-benzene [88], 5,5′-Methylenedifurfurylamine (DFDA) and 5,5′-Ethylidenedifurfurylamine (CH_3-DFDA) [89], diglycidyl ester of 2,5-furandicarboxylic acid, 2,5-Furandicarboxyli acid, and bis(prop-2-enyl) furan-2,5-dicarboxylate [90]. The phenolic and polyphenolic epoxy resins include catechin-based (such as glycidyl ether of catechin [91], glycidyl ether of green tea extract [92], glycidyl ether of heat dried green tea extract, and glycidyl ether of freeze-dried green tea [83]) or gallic acid-based (such as gallic acid epoxidized, tannic acid epoxidized, vanillic acid epoxidized [93–95], tri- and tetra-glycidyl ethers of gallic acid [96], cardanol epoxidized [97], cardanol novolac epoxy [98], cardanol [99], and tannic acid [95]). The hard segment of lignin can provide stiffness as a lignin-cured copolymer, which can significantly affect the properties of the copolymer [78]. The curing agent allows the main adhesive to form a net or three-dimensional structure to increase the cohesive strength of the adhesive layer, which is commonly formed by acid anhydrides, amines, sulfurs, and macromolecules [100].

Figure 6 describes the main route of lignin epoxy resin (LER) production by steam explosion-based biorefinery process. The steam-exploded substrate is subjected to the extraction process by water–methanol/acetone [4, 7, 8, 82] or direct methanol/acetone extraction under various temperatures and extraction times for water extraction. The extraction process provides LML after the evaporation process. The next step is resin synthesis using LML, which can be performed using various methods, such as a two-step reaction: epichlorohydrin catalyzed by tetrabutylammonium bromide (TBAB) to open the epoxy ring and the reaction with sodium hydroxide to reconstruct the hydrogen chloride [82]; lignin is reacted with epichlorohydrin and NaOH aqueous solution is added at 110°C, followed by the washing process, to deprive the NaOH to obtain LER [4]; lignin is reacted with epichlorohydrin and continue to use tetramethylammonium chloride (TMAC) by heated and flowed under nitrogen (N_2). This reaction is simpler than TBAB in order to syntheses the epoxy resin, and continue to epoxy ring reconstructed, after that the methyl ethyl ketone (MEK) solvent was used during the ring closure reaction and also for wash out the NaCl as by product of ring closure reaction [7, 8]. The curing process uses various curing agents, such as biological curing agents, epoxy resin, or chemical curing agent, to obtain cured LER through various processes. The cured process includes various biological or conventional curing agents.

LER synthesized from bamboo by steam explosion pretreatment was used for toxicity examination with human breast cancer estrogen-sensitive MCF cells. Here, the LER was synthesized from methanol-soluble lignin/LML and subjected to the epoxy reaction by melting the LML with a combination of epichlorohydrin and NAOH; then, the NaCl was produced from water extraction and evaporated to obtain epoxidized lignin. The LER production was continued by reacting with dimethyltyramine used as the curing agent. The epoxidized lignin is considered safe

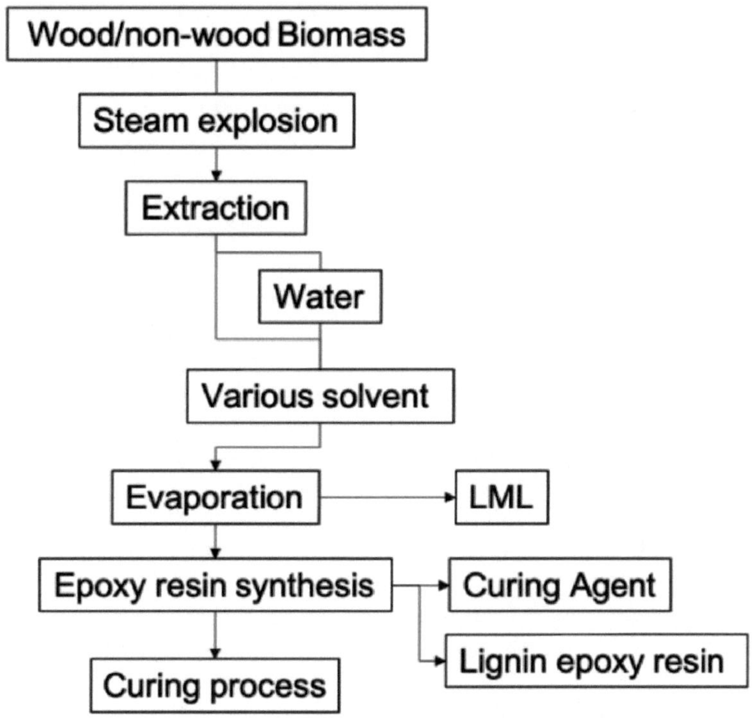

Figure 6.
Main routes for the conversion of lignocellulose to biopolymers by steam explosion pretreatment.

for the toxicity effect of epoxy resin because no proliferative MCF cells are produced, which indicates the absence of endocrine-disruption activity [4]. The epoxy resin syntheses produced an LER, which can act as an epoxy resin, a curing agent, or a copolymer to produce cured epoxy resin. Sasaki [82] introduced an LER that can be used as both an epoxy resin and a curing agent, and used epoxidized lignin resin from bamboo produced through a two-step reaction with catalysis transfer method by adding bamboo lignin to TBAB-catalyzed epichlorohydrin and releasing the hydrogen chloride to reconstruct the epoxy ring in the presence of sodium hydroxide. Asada [7] produced an LER that can be used as both a curing agent and an epoxy resin, which was evaluated by the synthesis of cured epoxy resin. The LER was synthesized by dissolving LML, followed by a catalyzed process, to open the epoxy ring and detaching the hydrogen chloride used TMAC. After the separation process, continue to be dissolved with MEK as a solvent to obtain LER by the washing process and vacuum-drying.

7. Antioxidant resources

LCC was broken down and degraded by steam explosion. The degradation of lignin [101, 102] and its compounds [103] resulted in a complete conversion of lignin to phenolic compounds, which is also an important raw material for the production of antioxidants and drugs [101]. The depolymerization was performed through the rupture of ether and destruction of C – C bonds connecting the phenylpropane units. This method produced low-molecular-weight monomer phenols, such as 2-methoxyphenols (guaiacol, 4-methylguaiacol, 4-ethylguaiacol, 4-vinylguaiacol, 4-propylguaiacol, eugenol, isoeugenol, vanillin, acetovanillone, and 2-propiovanillone),

2,6-dimethoxyphenols (syringol, 4-methylsyringol, 4-allylsyringol, syringaldehyde, and acetosyringone), dihydroxybenzenes (catechol, 3-methylcatechol, 4-methyl-catechol, 3-methoxycatecho, and hydroquinone), and phenolsyringaldehyde [6, 104–110], which are comparatively easy to identify through the separation and chromatographic identification [111].

Steam explosion pretreatment for antioxidant examination for biomass has been extensively reported. Asada [8] subjected white poplar to steam explosion pretreatment on a total biorefinery system and obtained 76 mg-catechin equiv./g-dry steam-exploded white poplar. Kurosumi [104, 112] used *S. palmata* from the leaf, stem, rhizome, and root of bamboo grass to examine its antioxidant compounds by steam explosion pretreatment followed by hot water extraction at 98°C for 2 h and methanol extraction. They examined the antioxidant activity of water-soluble material, methanol-soluble material, and its residue, which yielded a higher phenolic compound with a concentration of 217.41 mg/g from the leaf part, and expressed the antioxidant activity through butylated hydroxyanisole, which yielded 142.81 mg/g of radical scavenging activity from the leaf part. Noda [113] subjected raw garlic to steam explosion followed by water extraction to examine its antioxidant activity and obtained 0.135 g/l of EC_{50} with 80.8 mg/g of phenolic compounds. Subsequently, Noda [107] examined the garlic husk in comparison steam explosion with microwave irradiation followed by water extraction, and found that the value of EC_{50} was decreased, which was expressed as an increase in the radical scavenging activity. This is in line with the dramatically increased amount of phenolic compounds. Sui [114] examined the tea waste by steam explosion and observed a 20% improvement in the antioxidant capacity and OH, O_2, and ferric reducing antioxidant power (FRAP) radical scavenging activity of tea extracts. Romero [115] used olive leaves for obtaining the antioxidant by steam explosion pretreatment, and obtained 1950 mg of antioxidant from 100 g of olive leaves. Gong [116] used barley bran to determine the phenolic compounds and antioxidant activities with steam explosion followed by methanol extraction, and as a result, the total soluble phenolic content of 1686.4 gallic acid equivalents mg/100 g with 2983 TEAC mg/100 g of DW was obtained by the 2,2′-azinobis(3-ethylbenzothiazoline-6-sulfonic acid method for scavenging activity and 13.45 mmol $FeSO_4 \cdot L^{-1} \cdot g^{-1}$ was obtained using FRAP. Chen [117] used steam-exploded wheat to examine the antioxidant activity and antiproliferation on HepG2 cells by ultrasonic extraction with 80% methanol, followed by washing with phosphate buffer solution. They obtained 423.335 ± 19.94 mg/ml EC_{50} and concluded that the cellular and antiproliferation activities were enhanced. Furthermore, Li et al. [110] subjected the bran of buckwheat to steam explosion followed by the extraction process using 80% chilled acetone and a purification process to remove any lipid and other fractions. They examined the phenolic composition, antioxidant activity, and ability to inhibit the proliferation of HepG2 and Coca-2 cells, as well as their cytotoxicity, which resulted in 28.32 ± 0.91 and 13.18 ± 0.81 mg RE/g DW total phenolic and flavonoid, respectively. The antioxidant activity was expressed by oxygen radical absorbance capacity, which resulted in 1120.33 ± 41.43 μmol TE/g DW, and it can against the HepG2 and Caco-2 cells by antiproliferative activity. In addition, the effect of antioxidant extraction for ethanol production under the effect of saccharification and fermentation process has also been reported (e.g., [118–120]), which suggests that the scavenging activity cannot increase with the phenolic compounds. Asada [6] subjected softwood to steam explosion followed by water and methanol extraction. The water-soluble material obtained from water extraction as an antioxidant resource yielded 10.4 ± 2.52 min/μg/ml (AAPH-induced linoleic acid oxidation). This value is one-fifth that of Trolox, where the expectation was that the polyphenol compounds, such as phenylpropanoids and flavonoids, contained in water-soluble materials exhibit an antioxidative activity equal to that of Trolox. However, the result was below expectation, which is attributed to the impurities, and thus, the softwood must be purified before being examined for antioxidant activities.

8. Conclusions

The steam explosion pretreatment for lignocellulosic biomass biorefinery is an environmentally friendly process and can produce a wide range of products. This system aims to obtain an effective biorefinery to identify potentialities used in all sectors, including biomedical, aviation, energy, electronics, and environment. The previously reported studies on this topic have already provided several biorefinery systems based on steam explosion pretreatment for lignocellulosic biomass. Even though the scope is beyond that of the present review, it is necessary to perform the economical calculation and life cycle assessment analysis. Further studies must be conducted as most of the abovementioned lignocellulosic biomass byproducts have not yet been produced at a commercial production level and require a transition from academic assumptions to industry assumptions.

Acknowledgements

The authors gratefully acknowledge the financial supports provided by a Grant-in-Aid for Young Scientists (A) (Grant No. 17H04717) and a Grant-in-Aid for Scientific Research (A) (Grant No. 20H00664) from the Ministry of Education, Culture, Sports, Science, and Technology of Japan.

Author details

Chikako Asada[1], Sholahuddin[2] and Yoshitoshi Nakamura[1*]

1 Department of Bioscience and Bioindustry, Tokushima University, Tokushima, Japan

2 Graduated School of Life and Material System Engineering, Tokushima University, Tokushima, Japan

*Address all correspondence to: ynakamu@tokushima-u.ac.jp

References

[1] Mason WH. Process and apparatus for disintegration of wood and the like. US: US Patent; 1586159, 1926. p. 6.

[2] WH M. appartus for and process of explosion fibration of lignocellulose material. US; 1655618, 1928.

[3] Babcok L. Method for producing fermentable sugars and alcohol from wood. US; 1855464, 1932.

[4] Asada C, Nakamura Y, Kobayashi F. Waste reduction system for production of useful materials from un-utilized bamboo using steam explosion followed by various conversion methods. Biochemical Engineering Journal. 2005;23(2):131-137.

[5] Hongzhang C, Liying L. Unpolluted fractionation of wheat straw by steam explosion and ethanol extraction. Bioresource Technology. 2007;98(3):666-676.

[6] Asada C, Sasaki C, Uto Y, Sakafuji J, Nakamura Y. Effect of steam explosion pretreatment with ultra-high temperature and pressure on effective utilization of softwood biomass. Biochemical Engineering Journal. 2012;60:25-29. Available from: http://dx.doi.org/10.1016/j.bej.2011.09.013

[7] Asada C, Basnet S, Otsuka M, Sasaki C, Nakamura Y. Epoxy resin synthesis using low molecular weight lignin separated from various lignocellulosic materials. International Journal of Biological Macromolecules. 2015;74:413-419. Available from: http://dx.doi.org/10.1016/j.ijbiomac.2014.12.039

[8] Asada C, Sasaki C, Suzuki A, Nakamura Y. Total Biorefinery Process of Lignocellulosic Waste Using Steam Explosion Followed by Water and Acetone Extractions. Waste and Biomass Valorization. 2018;9(12):2423-2432. Available from: http://dx.doi.org/10.1007/s12649-017-0157-x

[9] Nakamura Y, Sawada T, Inoue E. Enhanced ethanol production from enzymatically treated steam-exploded rice straw using extractive fermentation. Journal of Chemical Technology and Biotechnology. 2001;76(8):879-884.

[10] Sasaki C, Kushiki Y, Asada C, Nakamura Y. Acetone-butanol-ethanol production by separate hydrolysis and fermentation (SHF) and simultaneous saccharification and fermentation (SSF) methods using acorns and wood chips of Quercus acutissima as a carbon source. Industrial Crops and Products. 2014;62:286-292. Available from: http://dx.doi.org/10.1016/j.indcrop.2014.08.049

[11] Sasaki C, Okumura R, Asada C, Nakamura Y. Steam explosion treatment for ethanol production from branches pruned from pear trees by simultaneous saccharification and fermentation. Bioscience, Biotechnology and Biochemistry. 2014;78(1):160-166.

[12] Kobayashi F, Sawada T, Nakamura Y, Ohnaga M, Godliving M, Ushiyama T. Saccharification and alcohol fermentation in starch solution of steam- exploded potato. Applied Biochemistry and Biotechnology - Part A Enzyme Engineering and Biotechnology. 1998;69(3):177-189.

[13] Nakamura Y, Sawada T. Ethanol production from artificial domestic household waste solubilized by steam explosion. Biotechnology and Bioprocess Engineering. 2003;8(3):205-209.

[14] Yamashita Y, Sasaki C, Nakamura Y. Effective enzyme saccharification and ethanol production from Japanese cedar using various pretreatment methods. Journal of Bioscience and

Bioengineering. 2010;110(1):79-86. Available from: http://dx.doi.org/10.1016/j.jbiosc.2009.12.009

[15] Asada C, Kita A, Sasaki C, Nakamura Y. Ethanol production from disposable aspen chopsticks using delignification pretreatments. Carbohydrate Polymers. 2011;85(1):196-200. Available from: http://dx.doi.org/10.1016/j.carbpol.2011.02.020

[16] Asada C, Sasaki C, Nakamura Y. High Concentration Ethanol Production from Mixed Softwood Sawdust Waste. Waste and Biomass Valorization. 2019;10(2):433-439.

[17] Asada C, Sasaki C, Takamatsu T, Nakamura Y. Conversion of steam-exploded cedar into ethanol using simultaneous saccharification, fermentation and detoxification process. Bioresource Technology. 2015;176:203-209. Available from: http://dx.doi.org/10.1016/j.biortech.2014.11.039

[18] Jurado M, Prieto A, Martínez-Alcalá Á, Martínez ÁT, Martínez MJ. Laccase detoxification of steam-exploded wheat straw for second generation bioethanol. Bioresource Technology. 2009;100(24):6378-6384. Available from: http://dx.doi.org/10.1016/j.biortech.2009.07.049

[19] Asada C, Asakawa A, Sasaki C, Nakamura Y. Characterization of the steam-exploded spent Shiitake mushroom medium and its efficient conversion to ethanol. Bioresource Technology. 2011;102(21):10052-10056. Available from: http://dx.doi.org/10.1016/j.biortech.2011.08.020

[20] Scholl AL, Menegol D, Pitarelo AP, Fontana RC, Filho AZ, Ramos LP, et al. Ethanol production from sugars obtained during enzymatic hydrolysis of elephant grass (Pennisetum purpureum, Schum.) pretreated by steam explosion. Bioresource Technology. 2015;192:228-237.

[21] Boluda-Aguilar M, López-Gómez A. Production of bioethanol by fermentation of lemon (*Citrus limon* L.) peel wastes pretreated with steam explosion. Industrial Crops and Products. 2013;41(1):188-97. Available from: http://dx.doi.org/10.1016/j.indcrop.2012.04.031

[22] Deng J, Zhu X, Chen P, He B, Tang SW, Zhao W, et al. Mechanism of lignocellulose modification and enzyme disadsorption for complete biomass saccharification to maximize bioethanol yield in rapeseed stalks. Sustainable Energy and Fuels. 2020;4(2):607-618.

[23] Asada C, Nakamura Y, Kobayashi F. Chemical characteristics and ethanol fermentation of the cellulose component in autohydrolyzed Bagasse. Biotechnol. Biopro. Eng. 2005;10(4):346-352.

[24] Ando S, Arai I, Kiyoto K, Hanai S. Identification of aromatic monomers in steam-exploded poplar and their influences on ethanol fermentation by Saccharomyces cerevisiae. Journal of Fermentation Technology. 1986;64(6):567-570.

[25] Anukam A, Mohammadi A, Naqvi M, Granström K. A Review of the Chemistry of Anaerobic Digestion: Methods of Accelerating and Optimizing Process Efficiency. Processes. 2019;7(8):504.

[26] Kotsyurbenko OR, Chin KJ, Glagolev M v., Stubner S, Simankova M v., Nozhevnikova AN, et al. Acetoclastic and hydrogenotrophic methane production and methanogenic populations in an acidic West-Siberian peat bog. Environmental Microbiology. 2004;6(11):1159-1173.

[27] Take H, Andou Y, Nakamura Y, Kobayashi F, Kurimoto Y, Kuwahara M. Production of methane gas from Japanese cedar chips pretreated by various delignification methods. Biochemical Engineering Journal. 2006;28(1):30-35.

[28] Kobayashi F, Take H, Asada C, Nakamura Y. Methane production from steam-exploded bamboo. Journal of Bioscience and Bioengineering. 2004;97(6):426-428.

[29] Mulat DG, Huerta SG, Kalyani D, Horn SJ. Enhancing methane production from lignocellulosic biomass by combined steam-explosion pretreatment and bioaugmentation with cellulolytic bacterium Caldicellulosiruptor bescii. Biotechnology for Biofuels. 2018;11(1):1-15. Available from: https://doi.org/10.1186/s13068-018-1025-z

[30] Sholahuddin, Nakamura Y, Asada C. Effect of Activated Cow Dung as Inoculum on Methane Production of Steam-Exploded Rice Husks. Waste and Biomass Valorization. 2021; Available from: https://doi.org/10.1007/s12649-021-01365-6

[31] Lizasoain J, Rincón M, Theuretzbacher F, Enguídanos R, Nielsen PJ, Potthast A, et al. Biogas production from reed biomass: Effect of pretreatment using different steam explosion conditions. Biomass and Bioenergy. 2016;95:84-91.

[32] Dererie DY, Trobro S, Momeni MH, Hansson H, Blomqvist J, Passoth V, et al. Improved bio-energy yields via sequential ethanol fermentation and biogas digestion of steam exploded oat straw. Bioresource Technology. 2011;102(6):4449-4455. Available from: http://dx.doi.org/10.1016/j.biortech.2010.12.096

[33] Yue ZB, Liu RH, Yu HQ, Cheng HZ, Yu B, Harada H, et al. Enhanced anaerobic ruminai degradation of bulrush through steam explosion pretreatment. Industrial and Engineering Chemistry Research. 2008;47(16):5899-5905.

[34] Fuchs G, Boll M, Heider J. Microbial degradation of aromatic compounds- From one strategy to four. Nature Reviews Microbiology. 2011;9(11):803-816. Available from: http://dx.doi.org/10.1038/nrmicro2652

[35] Yan L, Yue Z, Yu H. Microscale Analysis of in Vitro Anaerobic Degradation of Lignocellulosic Wastes by Rumen Microorganisms Microscale Analysis of in Vitro Anaerobic Degradation of Lignocellulosic Wastes by Rumen Microorganisms. 2015;42:276-281.

[36] Ko JJ, Shimizu Y, Ikeda K, Kim SK, Park CH, Matsui S. Biodegradation of high molecular weight lignin under sulfate reducing conditions: Lignin degradability and degradation by-products. Bioresource Technology. 2009;100(4):1622-1627. Available from: http://dx.doi.org/10.1016/j.biortech.2008.09.029

[37] Chen F, Dixon RA. Lignin modification improves fermentable sugar yields for biofuel production. Nature Biotechnology. 2007;25(7):759-761.

[38] Vivekanand V, Olsen EF, Eijsink VGH, Horn SJ. Effect of different steam explosion conditions on methane potential and enzymatic saccharification of birch. Bioresource Technology. 2013;127:343-349. Available from: http://dx.doi.org/10.1016/j.biortech.2012.09.118

[39] Healy JB, Young LY. Anaerobic biodegradation of eleven aromatic compounds to methane. Applied and Environmental Microbiology. 1979;38(1):84-89.

[40] Barakat A, Monlau F, Steyer JP, Carrere H. Effect of lignin-derived and furan compounds found in lignocellulosic hydrolysates on biomethane production. Bioresource Technology. 2012;104:90-99. Available from: http://dx.doi.org/10.1016/j.biortech.2011.10.060

[41] Fernández-Bolaños J, Felizón B, Heredia A, Rodríguez R, Guillén R, Jiménez A. Steam-explosion of olive stones: Hemicellulose solubilization and enhancement of enzymatic hydrolysis of cellulose. Bioresource Technology. 2001;79(1):53-61.

[42] Dai NH, Huynh KTT, Nguyen TAD, Do VVT, van Tran M. Hydrothermal and Steam Explosion Pretreatment of Bambusa stenostachya Bamboo. Waste and Biomass Valorization. 2020;(0123456789). Available from: https://doi.org/10.1007/s12649-020-01299-5

[43] Sasaki C, Okumura R, Asakawa A, Asada C, Nakamura Y. Effects of washing with water on enzymatic saccharification and d-lactic acid production from steam-exploded sugarcane bagasse. Journal of Material Cycles and Waste Management. 2012;14(3):234-240.

[44] Marcos M, García-Cubero MT, González-Benito G, Coca M, Bolado S, Lucas S. Optimization of the enzymatic hydrolysis conditions of steam-exploded wheat straw for maximum glucose and xylose recovery. Journal of Chemical Technology and Biotechnology. 2013;88(2):237-246.

[45] Liu ZH, Chen HZ. Simultaneous saccharification and co-fermentation for improving the xylose utilization of steam exploded corn stover at high solid loading. Bioresource Technology. 2016;201:15-26. Available from: http://dx.doi.org/10.1016/j.biortech.2015.11.023

[46] Wang W, Ling H, Zhao H. Steam explosion pretreatment of corn straw on xylose recovery and xylitol production using hydrolysate without detoxification. Process Biochemistry. 2015;50(10):1623-1628. Available from: http://dx.doi.org/10.1016/j.procbio.2015.06.001

[47] Nogi M, Yano H. Transparent nanocomposites based on cellulose produced by bacteria offer potential innovation in the electronics device industry. Advanced Materials. 2008;20(10):1849-1852.

[48] Fukuzumi H, Saito T, Iwata T, Kumamoto Y, Isogai A. Transparent and high gas barrier films of cellulose nanofibers prepared by TEMPO-mediated oxidation. Biomacromolecules. 2009;10(1):162-165.

[49] Xue Y, Mou Z, Xiao H. Nanocellulose as a sustainable biomass material: Structure, properties, present status and future prospects in biomedical applications. Nanoscale. 2017;9(39):14758-14781.

[50] Abdul Khalil HPS, Adnan AS, Yahya EB, Olaiya NG, Safrida S, Hossain MS, et al. A review on plant cellulose nanofibre-based aerogels for biomedical applications. Polymers. 2020;12(8).

[51] Amalraj A, Gopi S, Thomas S, Haponiuk JT. Cellulose Nanomaterials in Biomedical, Food, and Nutraceutical Applications: A Review. Macromolecular Symposia. 2018;380(1):1-9.

[52] Iwamoto S, Abe K, Yano H. The effect of hemicelluloses on wood pulp nanofibrillation and nanofiber network characteristics. Biomacromolecules. 2008;9(3):1022-1026.

[53] Mazzuckelli LF, Methner MM, Eileen Birch M, Evans DE, Ku BK, Crouch K, et al. Identification and characterization of potential sources of worker exposure to carbon nanofibers during polymer composite laboratory operations. Journal of Occupational and Environmental Hygiene. 2007;4(12):D125–D130.

[54] Saragih SW, Wirjosentono B, Eddiyanto, Meliana Y. Influence of

crosslinking agent on the morphology, chemical, crystallinity and thermal properties of cellulose nanofiber using steam explosion. Case Studies in Thermal Engineering. 2020;22:100740. Available from: https://doi.org/10.1016/j.csite.2020.100740

[55] Suzuki A, Sasaki C, Asada C, Nakamura Y. Production of cellulose nanofibers from Aspen and Bode chopsticks using a high temperature and high pressure steam treatment combined with milling. Carbohydrate Polymers. 2018;194(December 2017):303-10. Available from: https://doi.org/10.1016/j.carbpol.2018.04.047

[56] Dahlem MA, Borsoi C, Hansen B, Catto AL. Evaluation of different methods for extraction of nanocellulose from yerba mate residues. Carbohydrate Polymers. 2019;218:78-86. Available from: https://doi.org/10.1016/j.carbpol.2019.04.064

[57] Hongrattanavichit I, Aht-Ong D. Nanofibrillation and characterization of sugarcane bagasse agro-waste using water-based steam explosion and high-pressure homogenization. Journal of Cleaner Production. 2020;277:123471. Available from: https://doi.org/10.1016/j.jclepro.2020.123471

[58] Liu Q, Lu Y, Aguedo M, Jacquet N, Ouyang C, He W, et al. Isolation of High-Purity Cellulose Nanofibers from Wheat Straw through the Combined Environmentally Friendly Methods of Steam Explosion, Microwave-Assisted Hydrolysis, and Microfluidization. ACS Sustainable Chemistry and Engineering. 2017;5(7):6183-6191.

[59] Yang W, Cheng T, Feng Y, Qu J, He H, Yu X. Isolating cellulose nanofibers from steam-explosion pretreated corncobs using mild mechanochemical treatments. BioResources. 2017;12(4):9183-9197.

[60] Cherian BM, Leão AL, de Souza SF, Thomas S, Pothan LA, Kottaisamy M. Isolation of nanocellulose from pineapple leaf fibres by steam explosion. Carbohydrate Polymers. 2010;81(3):720-725.

[61] Anuradha Rani, Sheetal Monga, Monica Bansal AS. vm. Polymer Composites. 2016;00: 000-000:Thermoplastic starch (TPS)/cellulose nanofiber (CN.

[62] Yang W, Feng Y, He H, Yang Z. Environmentally-friendly extraction of cellulose nanofibers from steam-explosion pretreated sugar beet pulp. Materials. 2018;11(7).

[63] Deepa B, Abraham E, Cherian BM, Bismarck A, Blaker JJ, Pothan LA, et al. Structure, morphology and thermal characteristics of banana nano fibers obtained by steam explosion. Bioresource Technology. 2011;102(2):1988-1997. Available from: http://dx.doi.org/10.1016/j.biortech.2010.09.030

[64] Saragih SW, Wirjosentono B, Eddyanto, Meliana Y. Thermal and Morphological Properties of Cellulose Nanofiber from Pseudo-Stem Fiber of Abaca (Musa Textilis). Macromolecular Symposia. 2020;391(1):1-5.

[65] Supian MAF, Amin KNM, Jamari SS, Mohamad S. Production of cellulose nanofiber (CNF) from empty fruit bunch (EFB) via mechanical method. Journal of Environmental Chemical Engineering. 2020;8(1):103024. Available from: https://doi.org/10.1016/j.jece.2019.103024

[66] Kaushik A, Singh M, Verma G. Green nanocomposites based on thermoplastic starch and steam exploded cellulose nanofibrils from wheat straw. Carbohydrate Polymers. 2010;82(2):337-345. Available from: http://dx.doi.org/10.1016/j.carbpol.2010.04.063

[67] Feng YH, Cheng TY, Yang WG, Ma PT, He HZ, Yin XC, et al.

Characteristics and environmentally friendly extraction of cellulose nanofibrils from sugarcane bagasse. Industrial Crops and Products. 2018;111(March 2017):285-91. Available from: http://dx.doi.org/10.1016/j.indcrop.2017.10.041

[68] Wang Z, Qiao X, Sun K. Rice straw cellulose nanofibrils reinforced poly(vinyl alcohol) composite films. Carbohydrate Polymers. 2018;197(June):442-450.

[69] Suzuki A, Sasaki C, Asada C, Nakamura Y. Characterization of cellulose nanofiber from steam-exploded Japanese cedar. BioResources. 2017;12(4):7628-7641.

[70] Asada C, Seno M, Nakamura Y. Preparation of Biopolymer Composite Using Cedar-Derived Cellulose Nanofibers. Waste and Biomass Valorization. 2021; Available from: https://doi.org/10.1007/s12649-021-01436-8

[71] Asada C, Sasaki Y, Nakamura Y. Production of Eco-refinery Pulp from Moso Bamboo Using Steam Treatment Followed by Milling Treatment. Waste and Biomass Valorization. 2020;11(11):6139-6146. Available from: https://doi.org/10.1007/s12649-019-00847-y

[72] Xiao K, Deng J, Zeng L, Guo T, Gong Y, Yang B, et al. Enhancement of municipal sludge dewaterability by electrochemical pretreatment. Journal of Environmental Sciences (China). 2019;75:98-104. Available from: https://doi.org/10.1016/j.jes.2018.03.007

[73] Kaushik A, Singh M. Isolation and characterization of cellulose nanofibrils from wheat straw using steam explosion coupled with high shear homogenization. Carbohydrate Research. 2011;346(1):76-85. Available from: http://dx.doi.org/10.1016/j.carres.2010.10.020

[74] Panyasiri P, Yingkamhaeng N, Lam NT, Sukyai P. Extraction of cellulose nanofibrils from amylase-treated cassava bagasse using high-pressure homogenization. Cellulose. 2018;25(3):1757-1768. Available from: https://doi.org/10.1007/s10570-018-1686-6

[75] Kumari P, Pathak G, Gupta R, Sharma D, Meena A. Cellulose nanofibers from lignocellulosic biomass of lemongrass using enzymatic hydrolysis: characterization and cytotoxicity assessment. DARU, Journal of Pharmaceutical Sciences. 2019;27(2):683-693.

[76] Gea S, Siregar AH, Zaidar E, Harahap M, Indrawan DP, Perangin-Angin YA. Isolation and characterisation of cellulose nanofibre and lignin from oil palm empty fruit bunches. Materials. 2020;13(10).

[77] Shahi N, Min B, Sapkota B, Rangari VK. Eco-friendly cellulose nanofiber extraction from sugarcane bagasse and film fabrication. Sustainability (Switzerland). 2020;12(15):1-15.

[78] Asada C, Sholahuddin, Nakamura Y. Lignin as a Coating and Curing Agent on Biodegradable Epoxy Resins. In: Reactive and Functional Polymers Volume One. Springer International Publishing; 2020. p. 195-206.

[79] Ferdosian F, Yuan Z, Anderson M, Xu CC. Chemically modified lignin through epoxidation and its thermal properties. J-for. 2012;2(4):11-15.

[80] Ferdosian F, Yuan Z, Anderson M, Xu C (Charles). Synthesis and characterization of hydrolysis lignin-based epoxy resins. Industrial Crops and Products. 2016;91:295-301. Available from: http://dx.doi.org/10.1016/j.indcrop.2016.07.020

[81] Chikako Asada KH& YN. Utilization of Steam-Treated and Milling-Treated Lignin from Moso Bamboo as Curing Agent of Epoxy Resin. Waste and Biomass Valorization. 2021;

[82] Sasaki C, Wanaka M, Takagi H, Tamura S, Asada C, Nakamura Y. Evaluation of epoxy resins synthesized from steam-exploded bamboo lignin. Industrial Crops and Products. 2013;43(1):757-761. Available from: http://dx.doi.org/10.1016/j.indcrop.2012.08.018

[83] Basnet S, Otsuka M, Sasaki C, Asada C, Nakamura Y. Functionalization of the active ingredients of Japanese green tea (Camellia sinensis) for the synthesis of bio-based epoxy resin. Industrial Crops and Products. 2015;73:63-72. Available from: http://dx.doi.org/10.1016/j.indcrop.2015.03.091

[84] Nakamura Y, Sawada T, Kuno K, Nakamoto Y. Resinification by woody lignin and its characteristics on safety and biodegradation. Journal of Chemical Engineering of Japan. 2001;34(10):1309-1312.

[85] Fache M, Auvergne R, Boutevin B, Caillol S. New vanillin-derived diepoxy monomers for the synthesis of biobased thermosets. European Polymer Journal. 2015;67:527-538. Available from: http://dx.doi.org/10.1016/j.eurpolymj.2014.10.011

[86] Wang S, Ma S, Xu C, Liu Y, Dai J, Wang Z, et al. Vanillin-Derived High-Performance Flame Retardant Epoxy Resins: Facile Synthesis and Properties. Macromolecules. 2017;50(5):1892-1901.

[87] Cho JK, Lee JS, Jeong J, Kim B, Kim B, Kim S, et al. Synthesis of carbohydrate biomass-based furanic compounds bearing epoxide end group(s) and evaluation of their feasibility as adhesives. Journal of Adhesion Science and Technology. 2013;27(18-19):2127-2138.

[88] Hu F, la Scala JJ, Sadler JM, Palmese GR. Synthesis and characterization of thermosetting furan-based epoxy systems. Macromolecules. 2014;47(10):3332-3342.

[89] Hu F, Yadav SK, la Scala JJ, Sadler JM, Palmese GR. Preparation and Characterization of Fully Furan-Based Renewable Thermosetting Epoxy-Amine Systems. Macromolecular Chemistry and Physics. 2015;216(13):1441-1446.

[90] Deng J, Liu X, Li C, Jiang Y, Zhu J. Synthesis and properties of a bio-based epoxy resin from 2,5-furandicarboxylic acid (FDCA). RSC Advances. 2015;5(21):15930-15939. Available from: http://dx.doi.org/10.1039/C5RA00242G

[91] Nouailhas, H.; Aouf, C.; Le Guerneve, C.; Caillol, S.; Boutevin, B.; Fulcrand H. Synthesis and Properties of Biobased Epoxy Resins. Part 1. Glycidylation of Flavonoids by Epichlorohydrin. J Polym Sci Part A Polym Chem. 2011;49:2261-2270.

[92] Benyahya S, Aouf C, Caillol S, Boutevin B, Pascault JP, Fulcrand H. Functionalized green tea tannins as phenolic prepolymers for bio-based epoxy resins. Industrial Crops and Products. 2014;53:296-307. Available from: http://dx.doi.org/10.1016/j.indcrop.2013.12.045

[93] Aouf C, Lecomte J, Villeneuve P, Dubreucq E, Fulcrand H. Chemo-enzymatic functionalization of gallic and vanillic acids: Synthesis of bio-based epoxy resins prepolymers. Green Chemistry. 2012;14(8):2328-2336.

[94] Cao L, Liu X, Na H, Wu Y, Zheng W, Zhu J. How a bio-based epoxy monomer enhanced the properties of diglycidyl

ether of bisphenol A (DGEBA)/graphene composites. Journal of Materials Chemistry A. 2013;1(16):5081-5088.

[95] Mitsuhiro Shibata KN. Preparation and Properties of Biocomposites Composed of Bio-Based Epoxy Resin, Tannic Acid, and Microfibrillated Cellulose. Journal of Polymer Science: Part B: Polymer Physics. 2010;48:425-433.

[96] Antonella Tarzia, Jeanette Montanaro, Michele Casiello, Cosimo Annese, Angelo Nacci AM. Synthesis, Curing, and Properties of an Epoxy Resin Derived from Gallic Acid. BioResources. 2018;13(1):632-645.

[97] Unnikrishnan KP, Thachil ET. Synthesis and characterization of cardanol-based epoxy systems. Designed Monomers and Polymers. 2008;11(6):593-607.

[98] Atta AM, Al-Hodan HA, Hameed RSA, Ezzat AO. Preparation of green cardanol-based epoxy and hardener as primer coatings for petroleum and gas steel in marine environment. Progress in Organic Coatings. 2017;111(June):283-293. Available from: http://dx.doi.org/10.1016/j.porgcoat.2017.06.002

[99] Maffezzoli A, Calò E, Zurlo S, Mele G, Tarzia A, Stifani C. Cardanol based matrix biocomposites reinforced with natural fibres. Composites Science and Technology. 2004;64(6):839-845.

[100] Ren YL and S. Adhesives. In: Building Decorative Materials. 2011. p. 325-341.

[101] Wahyudiono, Sasaki M, Goto M. Recovery of phenolic compounds through the decomposition of lignin in near and supercritical water. Chemical Engineering and Processing: Process Intensification. 2008;47(9-10):1609-1619.

[102] Cheng L, Ye XP, He R, Liu S. Investigation of rapid conversion of switchgrass in subcritical water. Fuel Processing Technology. 2009;90(2):301-311. Available from: http://dx.doi.org/10.1016/j.fuproc.2008.09.009

[103] Wahyudiono, Sasaki M, Goto M. Conversion of biomass model compound under hydrothermal conditions using batch reactor. Fuel. 2009;88(9):1656-1664. Available from: http://dx.doi.org/10.1016/j.fuel.2009.02.028

[104] Kurosumi A, Sasaki C, Kumada K, Kobayashi F, Mtui G, Nakamura Y. Novel extraction method of antioxidant compounds from Sasa palmata (Bean) Nakai using steam explosion. Process Biochemistry. 2007;42(10):1449-1453.

[105] Noda Y, Asada C, Sasaki C, Hashimoto S, Nakamura Y. Extraction method for increasing antioxidant activity of raw garlic using steam explosion. Biochemical Engineering Journal. 2013;73:1-4. Available from: http://dx.doi.org/10.1016/j.bej.2013.01.013

[106] Asada C, Suzuki A, Nakamura Y. Antioxidant activity of water extract from bamboo by high-temperature and high-pressure steam treatment. Biomass Conversion and Biorefinery. 2021; Available from: https://doi.org/10.1007/s13399-021-01413-0

[107] Noda Y, Asada C, Sasaki C, Nakamura Y. Effects of Hydrothermal Methods such as Steam Explosion and Microwave Irradiation on Extraction of Water Soluble Antioxidant Materials from Garlic Husk. Waste and Biomass Valorization. 2019;10(11):3397-3402. Available from: http://dx.doi.org/10.1007/s12649-018-0353-3

[108] Bortolomeazzi R, Sebastianutto N, Toniolo R, Pizzariello A. Comparative evaluation of the antioxidant capacity of smoke flavouring phenols by crocin bleaching inhibition, DPPH radical

scavenging and oxidation potential. Food Chemistry. 2007;100(4): 1481-1489.

[109] Liu L, Zhao M, Liu X, Zhong K, Tong L, Zhou X, et al. Effect of steam explosion-assisted extraction on phenolic acid profiles and antioxidant properties of wheat bran. Journal of the science of food and agriculture. 2016;96(10):3484-3491.

[110] Li W, Zhang X, He X, Li F, Zhao J, Yin R, et al. Effects of steam explosion pretreatment on the composition and biological activities of tartary buckwheat bran phenolics. Food and Function. 2020;11(5):4648-4658.

[111] Kensingtort JKH, S N, Prague KM, Orinda LRS, Volumes S, Orinda EH. Analysis of phenols from lignin depolymerization by capillary gas chromatography. 1989;467(2).

[112] Kurosumi A, Kobayashi F, Nakamura Y. Development of New Extraction Method of Natural Antioxidants from Bamboo Grass. Transactions of the Materials Research Society of Japan. 2007;32(4):1139-1142.

[113] Noda Y, Asada C, Sasaki C, Hashimoto S, Nakamura Y. Extraction method for increasing antioxidant activity of raw garlic using steam explosion. Biochemical Engineering Journal. 2013;73:1-4. Available from: http://dx.doi.org/10.1016/j.bej.2013.01.013

[114] Sui W, Xiao Y, Liu R, Wu T, Zhang M. Steam explosion modification on tea waste to enhance bioactive compounds' extractability and antioxidant capacity of extracts. Journal of Food Engineering. 2019;261:51-9. Available from: https://doi.org:10.1016/j.jfoodeng.2019.03.015

[115] Romero-García JM, Lama-Muñoz A, Rodríguez-Gutiérrez G, Moya M, Ruiz E, Fernández-Bolaños J, et al. Obtaining sugars and natural antioxidants from olive leaves by steam-explosion. Food Chemistry. 2016;210:457-465.

[116] Gong L, Huang L, Zhang Y. Effect of steam explosion treatment on barley bran phenolic compounds and antioxidant capacity. Journal of Agricultural and Food Chemistry. 2012;60(29):7177-7184.

[117] Chen Y, Zhang R, Liu C, Zheng X, Liu B. Enhancing antioxidant activity and antiproliferation of wheat bran through steam flash explosion. Journal of Food Science and Technology. 2016;53(7):3028-3034.

[118] Cruz JM, Domínguez JM, Domínguez H, Parajó JC. Antioxidant and antimicrobial effects of extracts from hydrolysates of lignocellulosic materials. Journal of Agricultural and Food Chemistry. 2001;49(5): 2459-2464.

[119] Li J, Wu K, Xiao W, Zhang J, Lin J, Gong Y, et al. Effect of antioxidant extraction on the enzymatic hydrolysis and bioethanol production of the extracted steam-exploded sugarcane bagasse. Biochemical Engineering Journal. 2014;82:91-96.

[120] Li J, Lin J, Xiao W, Gong Y, Wang M, Zhou P, et al. Solvent extraction of antioxidants from steam exploded sugarcane bagasse and enzymatic convertibility of the solid fraction. Bioresource Technology. 2013;130:8-15.

Chapter 7

New Ethylenediamine Crosslinked 2D-Cellulose Adsorbent for Nanoencapsulation Removal of Pb (II) and Cu (II) Heavy Metal Ions: Synthesis, Characterization Application, and RSM-Modeling

Issam Jilal, Soufian El Barkany, Zahra Bahari, Youssef El Ouardi, Mohamed Loutou, Hassan Amhamdi, Mohamed Abou-Salama, Amin Salhi, Abderrahmane El Idrissi and Katri Laatikainen

Abstract

The main objective of the present work is to elaborate on a new eco-friendly and efficient adsorbent designated for aquatic micropollutants removal. However, the synthesis of the Ethylenediamine Crosslinked 2D-Cellulose green adsorbent was carried out successfully, by partial grafting of benzyl entities onto hydroxyl groups of HEC, and crosslinking with ethylenediamine ED. Further, the new ethylenediamine crosslinked 2D-Cellulose was used as a biosorbent for nanoencapsulation removal of copper and lead heavy metal ions from aqueous solutions. The proposal chemical structures of unmodified and modified materials were confirmed using FTIR, XRD, TGA, and SEM–EDX analysis. Furthermore, many parameters of the optimization for Pb (II) and Cu (II) in terms of removal efficiency including pH, adsorbent amount, and contact time were optimized by response surface methodology with a Box–Behnken design. Based on the desirability optimization with three factors, the maximal removal was 99.52% and 97.5% for Pb(II) and Cu (II), respectively and was obtained at pH = 5.94, 22.2 mg as the optimal adsorbent amount, and 21.53 min as contact time.

Keywords: Cellulose, Ethylenediamine, Adsorption, Lead, Copper, Nanoencapsulation

1. Introduction

In the last few decades, the fast development of industrialization and urbanization caused a tremendous and exponential increase in the human population, where the avoidance of the quality deterioration of water, air, and soil becomes the priority [1]. However, the pollution of aquatic environments by industrial effluents, in particular, water pollution by heavy metals, is the most serious problem due to their non-biodegradable properties, their persistence in the environmental media [2–6], toxicity to human health and ecosystems [7, 8]. Currently, the elimination of metal ions remains a major issue for environmental sustainability, where the greatest source of heavy metals contamination of wastewater is directly linked to anthropic activities, in particular industrial (textiles, rubber, leather, paper, plastic, coal, food, petrochemicals, etc.), agricultural (pesticides, forestry, etc.), pharmaceutical and hydrometallurgical activities [9–11]. Lead (Pb) is considered one of the main pollutants present in different components of the biosphere, where it accumulates through trophic chains or water intake [12]. In addition, Pb (II) ions are able, even at low concentrations, to causing severe central nervous system damage, kidney and immune system dysfunction in human beings, especially for children [8, 13–16]. Given its exceptional electrochemical aspect, copper occupies a prominent place in the international economy, where paving a large industrial area (energy, information, telecommunication, electronics, etc.), which is considered the second strategic raw material, in China, after oil [17, 18].

At the light of this, to decontaminate the industrial effluents from heavy metal ions, the efforts integrate the different separation techniques such as chemical precipitation [19, 20], membrane filtration [21, 22], flocculation and chemical coagulation [23, 24], biological treatment [25], ion exchange [26, 27], photocatalytic degradation [28], advanced oxidation [29], nanofiltration [30–32], and adsorption [33–38]. However, the adsorption process remains the most interesting and attractive technique due to its simplicity, ease of handling, high efficiency, ability to remove a large amount of organic and inorganic pollutants, and availability [39–42]. Faced with environmental constraints, the modern industrial policy requires well-defined qualities for basic adsorbent materials such as stability, retention capacity, low cost, biodegradability, etc. [43]. In this regard, cellulose, the most abundant, renewable, non-toxic, biodegradable, biocompatible, inexpensive, and environmentally friendly biopolymer in the world [44–48] has attracted increasing attention meeting ecological criteria [35, 49–51]. However, native cellulose showed a low adsorption capacity that indicating a low surface charge density. Thus, to improve its reactivity, it is necessary to reduce structurally related recalcitrance, where the exceptional chemical structure of cellulose, in particular the hydrogen bond type interactions at the supramolecular level, prevents its dissolution in almost all conventional solvents [52]. Another way to introduce new functionalities is to use cellulosic derivatives with more reactive accessibility. A good example of these derivatives is hydroxyethylcellulose (HEC), which makes it possible to obtain good solubility in an alkaline medium and water [53]. In addition, the reactivity of HEC is greater than that of cellulose because the density of the primary alcohol entities has increased [33], which gives the possibility of good control of the degree of substitution (DS), in particular for Williamson etherification of HEC grafted benzyl entities, where solubility has been successfully achieved and controlled in common solvents [54].

In this work, HEC was used as a water-soluble derivative of cellulose, where it was partially hydrophobized by introducing benzyl functionalities to decrease the average functionality of the reaction system, to avoid the 3D crosslinking in the following stage of the reaction, which ensures good accessibility of the pollutant load during the adsorption process, this ecological synthesis showed, in our previous article, that the degree of substitution does not exceed unity [33, 53–55]. Then, the benzoxyethylcellulose (BEC) polymer chains were crosslinked using ethylenediamine (ED) as a crosslinking agent and as a bidentate metal ion chelator, especially for the removal of Pb (II) or Cd (II) [56–64]. Structural analyzes showed good agreement with the proposed structure of the new green adsorbent (BEC-ED). X-ray diffraction patterns and SEM confirmed the emergence of a new structural order of BEC-ED at the crystalline and morphological levels. However, the elemental surface profile of BEC-ED has been proven based on the EDS spectra. The effect of crosslinking on the ability to remove heavy metals from aquatic environments was studied as a function of physicochemical parameters (pH, nature of ions, contact time, initial concentration, etc.), which are optimized using the response surface methodology with the *Box–Behnken* design. To our knowledge, based on a review of the literature, the new green BEC-ED has never been described previously.

2. Materials and methods

2.1 Materials

HEC (DS $\sim$ 1.5) of 95% purity was purchased from HIMEDIA and it was purified by the dissolution-precipitation method in water-acetone solvents. Tetrahydrofuran (THF), ethylenediamine ED, thionyl chloride ($SOCl_2$), DMSO, hydrochloric acid (HCl) 37%, and triethylamine $(Et)_3N$ were used as received from Sigma-Aldrich. Sodium hydroxide (NaOH) and Benzyl bromide were purchased from Merck. All other chemicals are analytical grade and were used as received without any further purification.

2.2 Methods

Fourier Transform Infrared spectroscopy (FTIR) spectra of the HEC, BEC, and BEC-ED samples were recorded on Shimadzu FTIR-8400S spectrometer using finely ground KBr pellets with 2% of the sample at a resolution of 2 cm^{-1}. The measurements were performed over from 4000 to 400 cm^{-1}, and averages of 40 scans were taken for each sample. The morphological SEM images of HEC, BEC, and BEC-ED were investigated using scanning electron microscopy (TESCAN VEGA 3 LM), with an accelerating voltage of 10 kV. Energy-dispersive X-ray spectra (EDS) were recorded to divulge the elemental profile presenting on unmodified and modified polymer surfaces. The sample crystal orders were evaluated using the X-ray diffraction technique and were obtained from EQUINOX 2000an X-ray Diffractometer, using copper radiation CuKα (λ = 1.5418 Å), at an accelerating voltage of 40 kV and an operating current of 30 mA. All patterns are recorded in the range of 2θ (5°–35°). 0.25 g of each sample was pressed under 50 MPa to form pellets having an average of 25 mm in diameter. The thermal behavior of each sample was carried out on simultaneous DTA-TG Shimadzu DTG-60 apparatus. The amounts of the samples were between 8 and 12 mg and the interval of the temperature measurement was between room temperature and 600°C with a heating rate of 10°C min^{-1} under nitrogen flow. For all experiments, lead and copper ions concentrations were obtained by inductively coupled plasma-atomic emission spectroscopy on an ICP-AES (Iris Intrepid IIXDL ICP-AES).

2.2.1 Preparation of BEC-ED

BEC was prepared according to the method described in our previous paper [54]. First, the crosslinking of BEC is carried out by passing through the chlorination of the free OH groups, as an intermediate step. Where, to 1 g (2.6 mmol) of the BEC dissolved in the minimum of DMSO, 0.38 ml (2.7 mmol) of $(Et)_3N$ and

0.40 ml (5.5 mmol) of Thionyl chloride $SOCl_2$ were added dropwise. The reaction mixture was heated at 70° C for 2 h under stirring. The product is precipitated in water, filtered under vacuum, and then treated with 2% ammonia solution to neutral pH. Finally, the sample was washed frequently with water, filtered, and dried at 70°C. Ethylenediamine crosslinked BEC (BEC-ED) was carried out in THF at reflux for 4 h, where 1 g of BEC-Cl was reacted with ED in large excess (6 ml) using TEA as a capturing agent of HCl released. At the end of the reaction, the resulting product (white powder) was isolated by filtration under vacuum and frequently laved by distilled water to remove the ammonium salt and ED execs.

3. Results and discussions

3.1 Synthesis and characterization

BEC-ED synthesis was performed after HEC hydrophobization (partial benzylation) to decrease the rate of crosslinking in bio-adsorbent. In addition, crosslinking was performed with ED to study the effect of graft grouping on the ability to remove heavy metals from aquatic environments. The originality of this work is summed up in the fact that this type of product has never been described previously in the literature. The reaction scheme for the synthesis of BEC-ED, as a novel heavy metal adsorbent in aquatic media, is shown in **Figure 1**. The

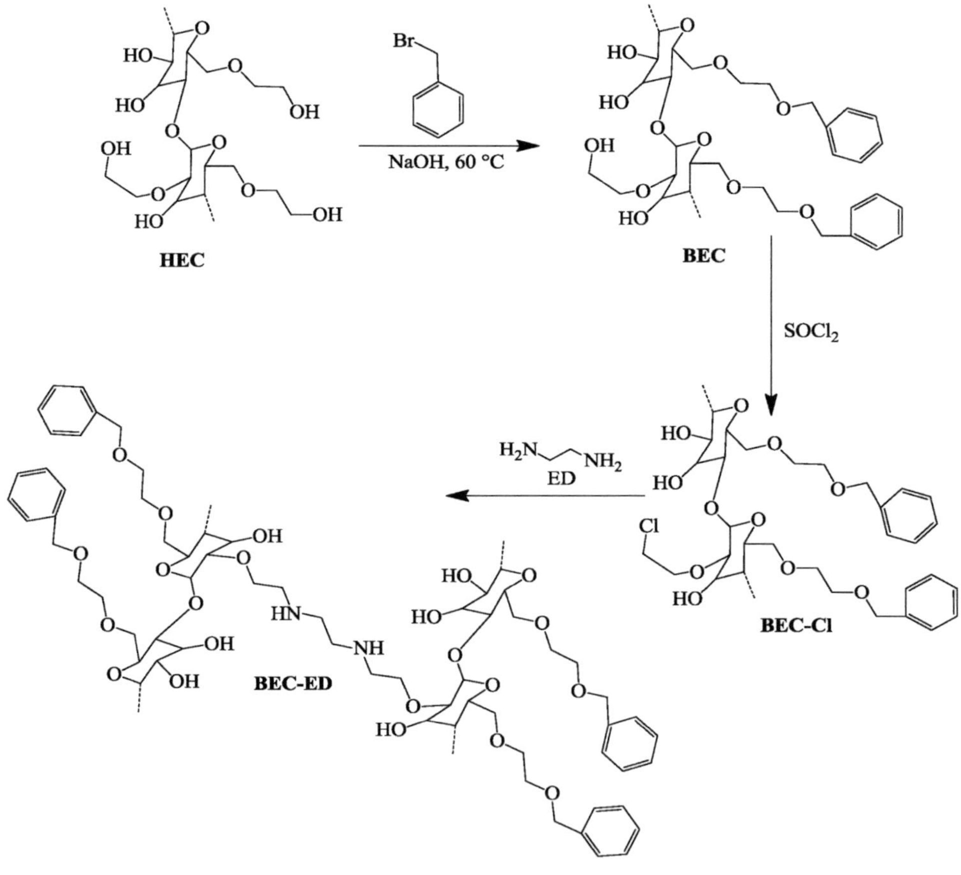

Figure 1.
Reaction scheme of the preparation of BEC-ED.

crosslinking of BEC is carried out through the chlorination of the free OH groups, as an intermediate step, and then crosslinked by the ED in the THF using Triethylamine as the capturing agent of the released HCl to avoid the degradation of the cellulose chain under the effect of acid. However, the apparition of the white powder during the reaction indicating that the reaction of the crosslinking has been carried out successfully.

3.1.1 Structural analysis (FTIR)

FTIR spectra of HEC, BEC, BEC-Cl, and BEC-ED are given in **Figure 2**. The FTIR of unmodified HEC spectrum showed infrared absorption bands spotted at 1062, 1408, 1458, 2873, 2927, and 3412 cm^{-1}. The absorption band at 3412 cm^{-1} is attributed to O–H stretching vibration [65], and a medium absorption band located in the range of 2927 and 2873 cm^{-1} corresponds to the C–H stretching vibration [66]. Moreover, the characteristic bands situated around 1408 and 1458 cm^{-1} are attributed to C–H symmetric bending vibration in –CHOH and O–H plane deformation of a primary alcohol, respectively [55]. The absorption band of b-(1,4)

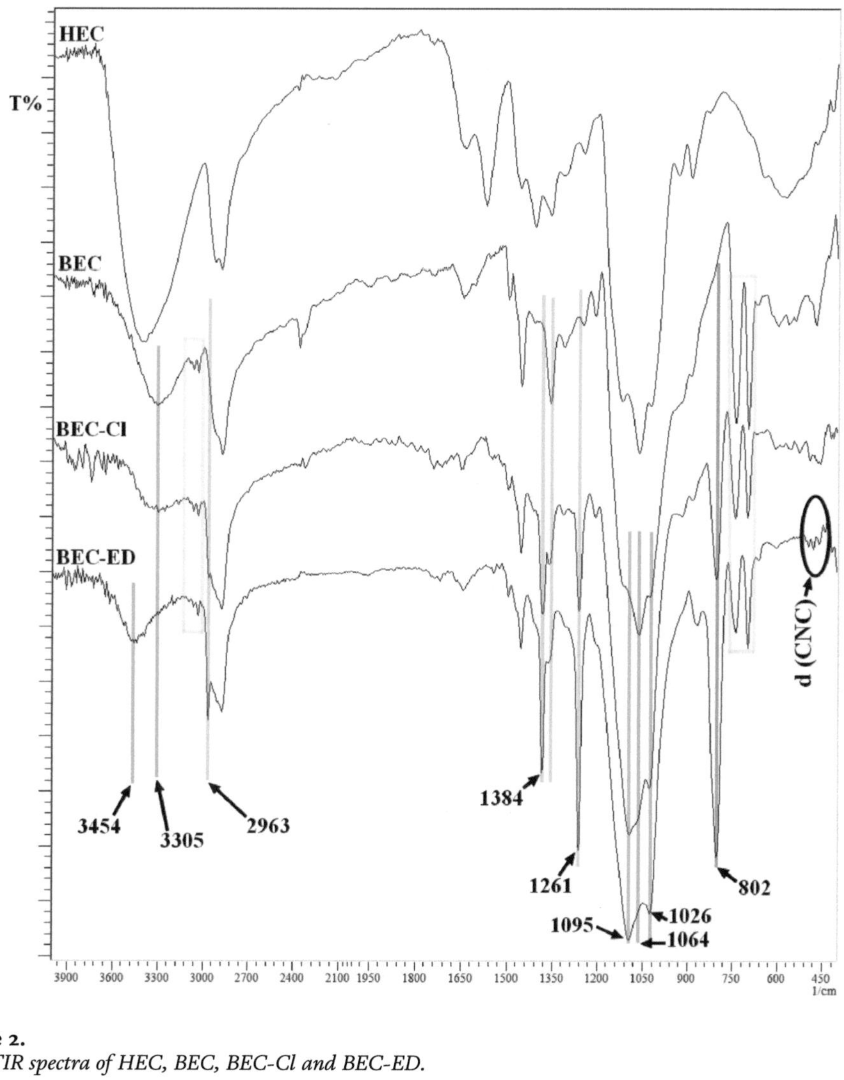

Figure 2.
The FTIR spectra of HEC, BEC, BEC-Cl and BEC-ED.

glycoside linkage was observed at 887 cm^{-1} [67], and that of C–O–C stretching vibration in the glucopyranose at 1062 cm^{-1} [68]. The absorption band at 1120 cm^{-1} corresponds to the C–O asymmetric vibration [67]. It can be seen, in **Figure 2**, that the modification of HEC by the benzyl group is apparent with a decrease in the intensity of the peak at 3347 cm^{-1} indicating a benzyl substitution of OH groups [69]. Indeed, the aromatic characteristic band elongations (=C–H) are situated between 3090 and 3033 cm^{-1} [70] and the aromatic C=Csp2 elongation vibrations are located at 1454 cm^{-1} [71]. In addition, the appearance of new absorption bands corresponding to the angular deformation (out of plane) of the monosubstituted aromatic C-H at around 740 cm^{-1} [72], and the C=C aromatic angular deformation, situated at 698 cm^{-1}, is a strong indication of the benzyl group incorporation on the HEC polymeric structure.

The comparison of FTIR spectra of unmodified BEC and chlorinated BEC (BEC-Cl) shows that the chlorination of BEC was carried out with success. Indeed, the new absorption band at 802 cm^{-1}, attributed to the stretching of the carbon-chlorine bond C-Cl, is a strong indication that confirming the chlorination reaction. In addition, the decrease in band intensity at 3305 cm^{-1} is due to the substitution of the hydroxyl group by chlorine, which confirms the success of the reaction [73, 74]. After BEC crosslinking, the appearance of the characteristic –NH– absorption band between 3305 cm^{-1} and 3454 cm^{-1} designates the incorporation of amino entities into the BEC structure. The increase in the density of $-CH_2-$ groups in the cellulosic skeleton is noticed through the increase in the intensity of the absorption band corresponding to the stretching vibrations of the methylene ($-CH_2$) groups at 2963 cm^{-1}. Furthermore, the ED crosslinking BEC is confirmed by the appearance of the different characteristic bands of the amino groups, which are located at 1095 cm^{-1} and 1590 cm^{-1} corresponding to the NH and CN stretching vibrations, respectively [75]. Thus, the intense peak attributed to the out-of-plane strain of NH at 802 cm^{-1} is very remarkable [75]. Also, the reduction in the intensity of the CO alcohol characteristic band around 1200 cm^{-1} is a strong indication of the substitution of OH by NH of ED [76–78].

3.1.2 Scanning electron microscopy, energy-dispersive x-ray (SEM-EDS) spectroscopy

Figure 3 shows SEM images of HEC, BEC, and BEC-ED. The resulted SEM images obtained for BEC showed homogenous, continuous, and microporous morphology, where pores diameter was estimated about (1–2) μm, which is radically different from the HEC aggregation aspect and lamellar BEC-ED morphologies. Yet, the morphological character of BEC allows it to be considered as a good candidate for microporous adsorbent/membranes applications. However, the EDS spectra of HEC and BEC showed a very significant increase in the C/O ratio, which indicates that the benzyl entities are grafted successfully. The evidence of BEC crosslinking by ED is shown by the EDS spectrum corresponding to BEC-ED, where the peak corresponding to nitrogen is very noticeable. In addition, BEC-ED SEM images showed a Nanoscale laminated appearance, including a lamellar structure that occurs through hydrophobic interactions. Based on BEC-ED morphology results, a supramolecular structure is proposed and schematically illustrated in **Figure 4**.

3.1.3 X-ray diffraction (XRD)

Figure 5 shows the X-ray diffractograms of HEC, BEC, and modified BEC (BEC-ED) in the range of 2θ = 00° to 37°. According to the diffractograms of the two polymers (HEC and BEC), the benzylation of HEC has practically no effect on the crystal behavior of HEC, except for a small shift of the maximum diffraction

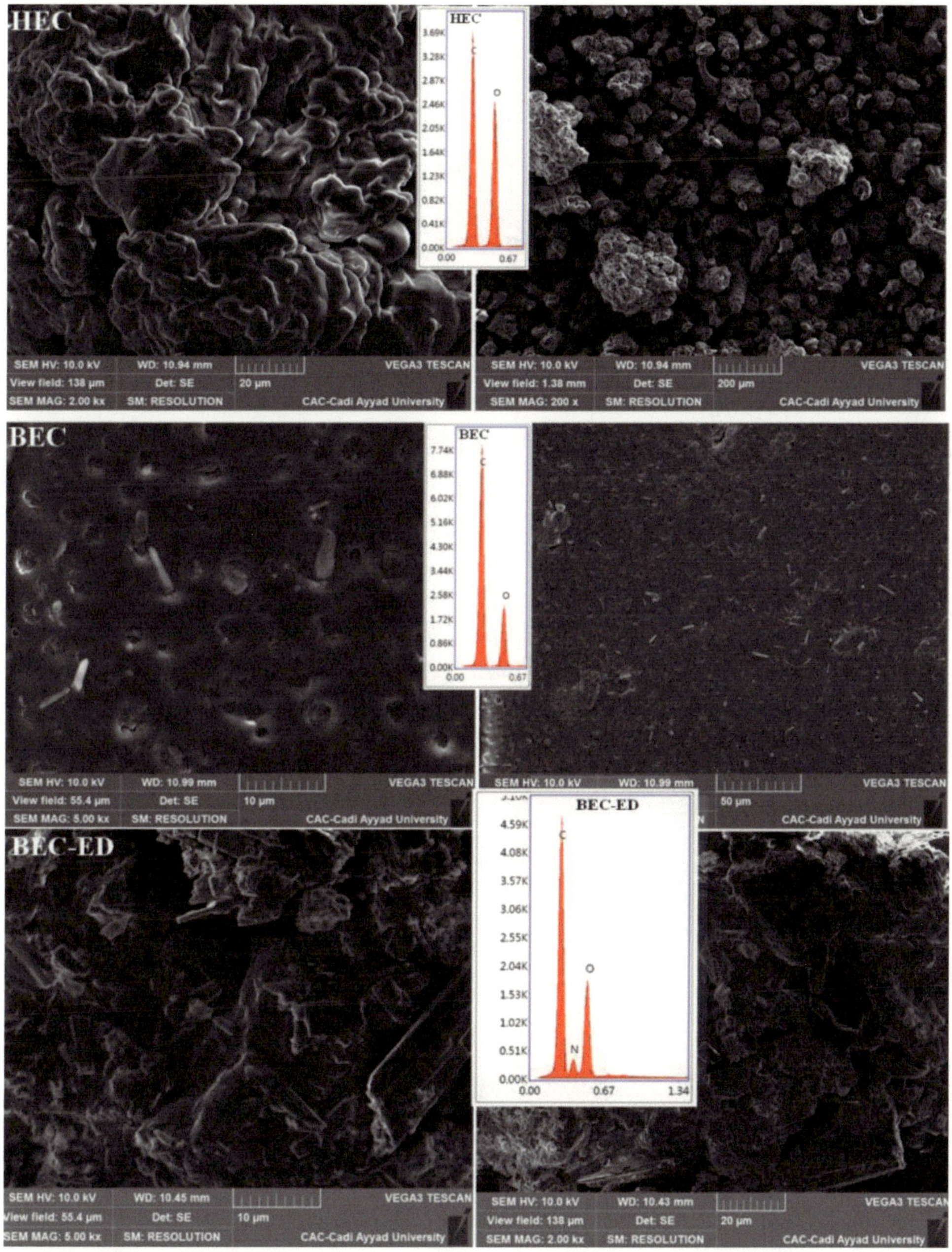

Figure 3.
SEM images and EDS spectra of HEC, BEC and BEC-ED.

peak towards the low values of HEC, 2θ of 21.25° for HEC and 20.80° for BEC, but the predominance of the amorphous character is always considered. Though, the small decrease in 2θ, at this region, can be explained by the increase in the supramolecular distance, between macromolecular chains, resulting from the insertion of benzyl entities. On the other hand, the diffractogram of the modified BEC sample (BEC-ED) showed, comparing to BEC, very remarkable and significant changes. Indeed, the crosslinking reaction of BEC generated a new crystalline order characterized by the apparition of a new peak towards 2θ = 12.50°, which corresponds to a lattice distance of 07.07 Å (**Figure 5**). The presence of a large domain with a

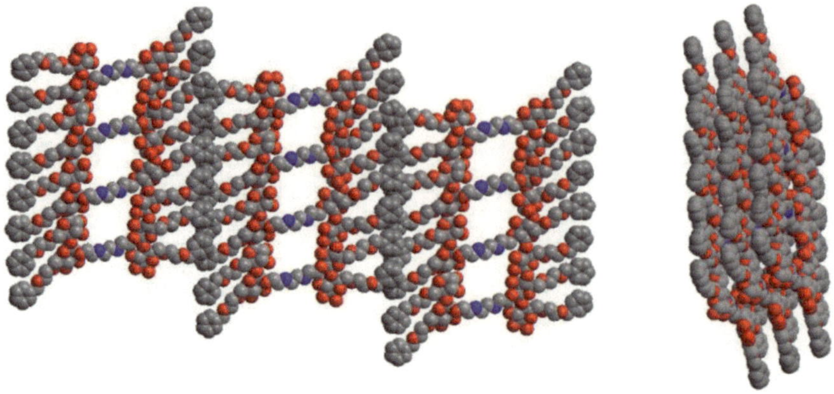

Figure 4.
Supramolecular lamellar structure of BEC-ED.

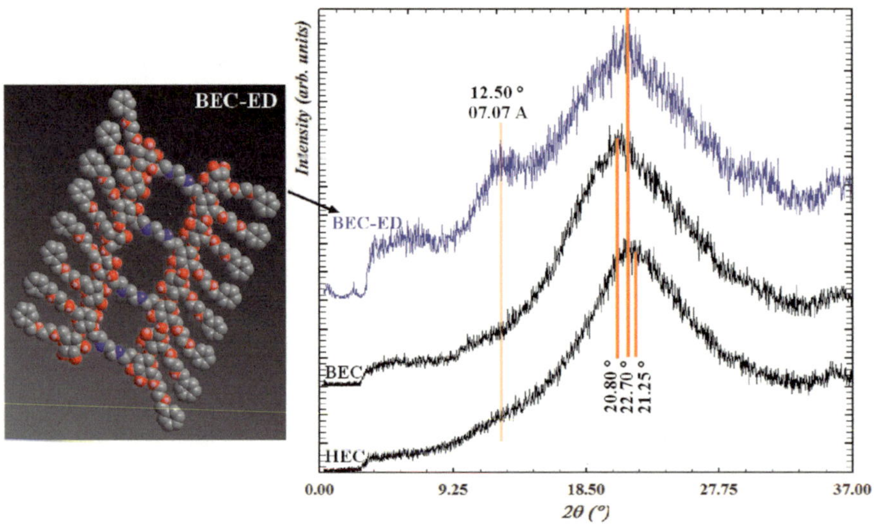

Figure 5.
Supramolecular structure of BEC-ED and the XRD diffractograms of HEC, BEC and BEC-ED.

maximum at 2θ = 22.7°, on the BEC-ED diffractogram, reveals its amorphous character. Therefore, a semi-crystalline appearance of BEC-ED can be suggested.

3.1.4 Thermogravimetric analysis (TGA)

Thermogravimetric makes it possible to follow, as a function of temperature, the weight loss evolution of each sample, mainly caused by dehydration or/and by the decomposition of the organic matter it contains. The thermal stability of BEC-ED comparing to BEC is studied basing on the TGA thermograms shown in **Figure 6**. The thermal behavior of BEC shows two stages of thermal decomposition, the first one is observed between 35 and 250°C attributed to the solvents and adsorbed water vaporization [79]. The strong weight loss (70%) corresponds to the degradation of the grafted entities and the cellulose backbone is observed in the temperature range of 250–600°C. For BEC-ED, no thermal event was observed below 250°C, indicating the absence of traces of solvents. The thermal decomposition of BEC-ED is noticed from 250°C up to 450°C with a mass loss of 90%. In addition, a low degradation of the thermal stability of BEC-ED compared to BEC has been noted,

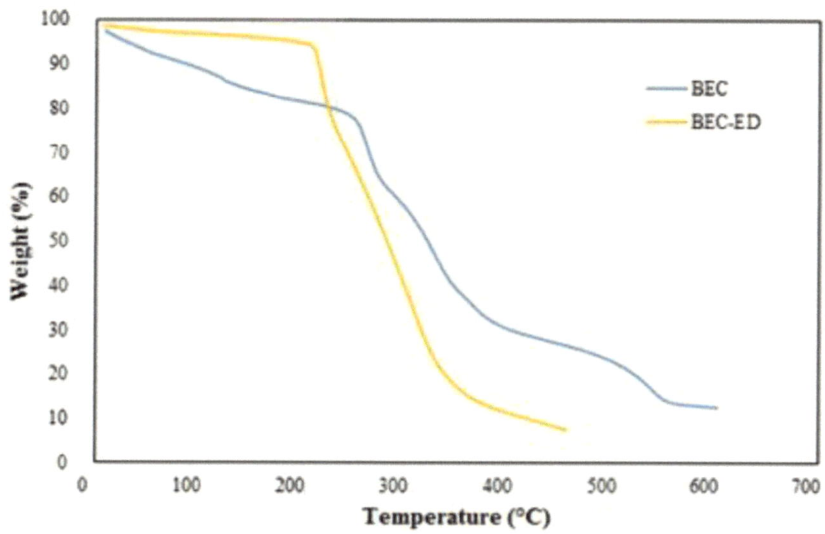

Figure 6.
TGA thermogram profiles of BEC and BEC-ED.

Variables	Factors	Unit	Level 1 (−1)	Level 2 (0)	Level 3 (1)
X_1	Ph	—	4.5	6	7.5
X_2	Adsorbent amount	mg	10	20	30
X_3	Contact time	min	5	17.5	30

Table 1.
Coded and actual variables and their levels.

and this is possibly due to the decrease of hydrogen interaction density and their replacement by hydrophobic interactions, and the supramolecular separation of the polymer chains caused by grafted ethylenediamine (ED).

3.2 Response surface methodology (RSM) modeling procedure

In the current work, is focused on *3-level Box–Behnken design* (BBD) in response surface methodology for seeking the optimal conditions for the removal efficiency of Cu(II) and Pb(II) onto BEC-ED. The three variables affecting the current process are pH at 4.5, 6.0, and 7.5, contact time at 5, 17.5, and 30 min, and adsorbent amount at 10, 20 and 30 mg. The complete design consisted of three different levels (−1, 0, and + 1) and 3-variable (pH—X_1, Adsorbent amount —X_2, and contact time —X_3). The layout of the factorial design is shown in **Table 1**. A total of 17 experiments were used in this study to evaluate the effects of the three input variables on Pb(II) and Cu(II) removal efficiency. The full picture of experiments with their responses (Pb(II) and Cu(II) removals) are tabulated in **Tables 2** and **3**, respectively.

The analysis of variance (ANOVA) was applied to the experimental runs, and then the results of the Box–Behnken design table are calculated and fitted by a suitable polynomial equation. According to the model's evaluation in **Tables 4** and **5**, which focuses on maximum R^2, predicted R^2, and adjusted R^2, the quadratic polynomial (Eq. (1)) model was chosen and well-fitted for all three independent parameters and responses (Cu(II) and Pb(II) removal efficiency).

	Factor 1	Factor 2	Factor 3	Response 1	Response 2
Run	X_1:pH	X_2:Adsorbent amount	X_3:Contact time	Pb removal	Cu removal
		Mg	min	%	%
1	0	0	0	95.64	91.21
2	1	-1	0	40.27	43.24
3	-1	1	0	61.85	57.57
4	0	0	0	96.41	93.32
5	-1	0	-1	43.53	45.92
6	-1	-1	0	42.85	39.84
7	0	0	0	96.23	94.78
8	0	1	-1	69.08	73.12
9	1	0	-1	42.82	51.48
10	0	1	1	92.42	91.37
11	1	1	0	62.32	61.76
12	0	-1	1	68.34	62.28
13	-1	0	1	67.24	68.84
14	0	0	0	94.86	94.81
15	0	-1	-1	47.17	49.53
16	0	0	0	93.81	91.76
17	1	0	1	64.87	65.67

Table 2.
The BBD matrix design with three independent factors and the corresponding experimental results.

Source	Sum of Squares	df	Mean Square	F-value	p-value	
Model	7460.88	9	828.99	730.00	< 0.0001	Significant
A-pH	3.37	1	3.37	2.96	0.1288	
B-Amount	947.00	1	947.00	833.92	< 0.0001	
C-Time	1018.58	1	1018.58	896.96	< 0.0001	
AB	2.33	1	2.33	2.05	0.1955	
AC	0.6889	1	0.6889	0.6066	0.4616	
BC	1.18	1	1.18	1.04	0.3425	
A^2	3566.13	1	3566.13	3140.31	< 0.0001	
B^2	880.99	1	880.99	775.80	< 0.0001	
C^2	573.67	1	573.67	505.17	< 0.0001	
Residual	7.95	7	1.14			
Lack of Fit	3.36	3	1.12	0.9779	0.4866	Not significant
Pure Error	4.59	4	1.15			
Cor Total	7468.83	16				

Table 3.
ANOVA analyses of the quadratic model and determination coefficients for Pb(II) adsorption efficiency.

Source	Std. Dev.	R^2	Adjusted R^2	Predicted R^2	PRESS	
Linear	20.57	0.2636	0.0937	−0.1046	8249.92	
2FI	23.44	0.2642	−0.1773	−0.9645	14672.20	
Quadratic	1.07	0.9989	0.9976	0.9918	60.98	Suggested
Cubic	1.07	0.9994	0.9975		*	Aliased

**Case(s) with leverage of 1.0000: PRESS statistic not defined.*

Table 4.
Model summary statistics Pb(II).

Source	Std. Dev.	R^2	Adjusted R^2	Predicted R^2	PRESS	
Linear	19.32	0.2466	0.0728	−0.1445	7370.22	
2FI	21.97	0.2508	−0.1987	−1.0514	13210.51	
Quadratic	2.73	0.9919	0.9815	0.8955	672.91	Suggested
Cubic	1.67	0.9983	0.9931		*	Aliased

**Case(s) with leverage of 1.0000: PRESS statistic not defined.*

Table 5.
Model summary statistics Cu(II).

Therefore, the predictive polynomial quadratic response model can be described as the following equation (Eq. (1)) [80]:

$$Y = \beta_0 + \sum_{i=1}^{n} \beta_i X_i + \sum_{i=1}^{n} \beta_{ii} X_i^2 + \sum_{i=1}^{n} \sum_{i>1}^{n} \beta_j X_i X_j \quad (1)$$

Where Y is the predicted response, β0 and βi are the intercept coefficient, and the linear coefficient respectively, βii and βij are the quadratic and the interaction coefficients, respectively, while Xi and Xj represent the coded values of the independent variables.

An ANOVA analysis for Cu (II) and Pb (II) removals was performed, and the results are presented in **Tables 3** and **6**, respectively. According to ANOVA analysis, the results obtained showed that the F and P-values less than 1000 and 0.0500, respectively. This confirmed that the model terms are significant. While **Lack of Fit F-value** in the ANOVA tables introduces an insignificant error with regard to the pure error. The response for Cu(II) and Pb(II) removal efficiency was determined with real factors by the following expressions (Eqs. (2) and (3)):

$$\begin{aligned} Pb(II)\ Removal &= -480.655 + 154.152 * pH + 6.49305 * Amount + 3.56334 * Time \\ &+ 0.0508333 * pH * Amount + -0.0221333 * pH * Time + 0.00434 * Amount * Time \\ &+ -12.9344 * pH^2 + -0.14465 * Adsorbent\ amount^2 + -0.074704 * Time^2 \end{aligned} \quad (2)$$

$$\begin{aligned} Cu(II)\ Removal &= -462.971 + 146.053 * pH + 7.18408 * Amount + 3.02441 * Time \\ &+ 0.00816667 * pH * Amount + -0.1164 * pH * Time + 0.011 * Amount * Time + \\ &-11.9436 * pH^2 + -0.157755 * Amount^2 + -0.0532832 * Time^2 \end{aligned} \quad (3)$$

Statistical diagnostics test is an excellent and effective tool for confirming the model presented. These diagnostic plots are given in **Figure 7**. By classifying the

Source	Sum of Squares	df	Mean Square	F-value	p-value	
Model	6387.74	9	709.75	95.35	< 0.0001	Significant
A-pH	13.21	1	13.21	1.77	0.2245	
B-Amount	995.25	1	995.25	133.71	< 0.0001	
C-Time	579.87	1	579.87	77.90	< 0.0001	
AB	0.0600	1	0.0600	0.0081	0.9310	
AC	19.05	1	19.05	2.56	0.1536	
BC	7.56	1	7.56	1.02	0.3470	
A^2	3040.67	1	3040.67	408.51	< 0.0001	
B^2	1047.86	1	1047.86	140.78	< 0.0001	
C^2	291.85	1	291.85	39.21	0.0004	
Residual	52.10	7	7.44			
Lack of Fit	40.97	3	13.66	4.91	0.0792	Not significant
Pure Error	11.13	4	2.78			
Cor Total	6439.84	16				

Table 6.
ANOVA analyses of the quadratic model and determination coefficients for Cu(II) adsorption efficiency.

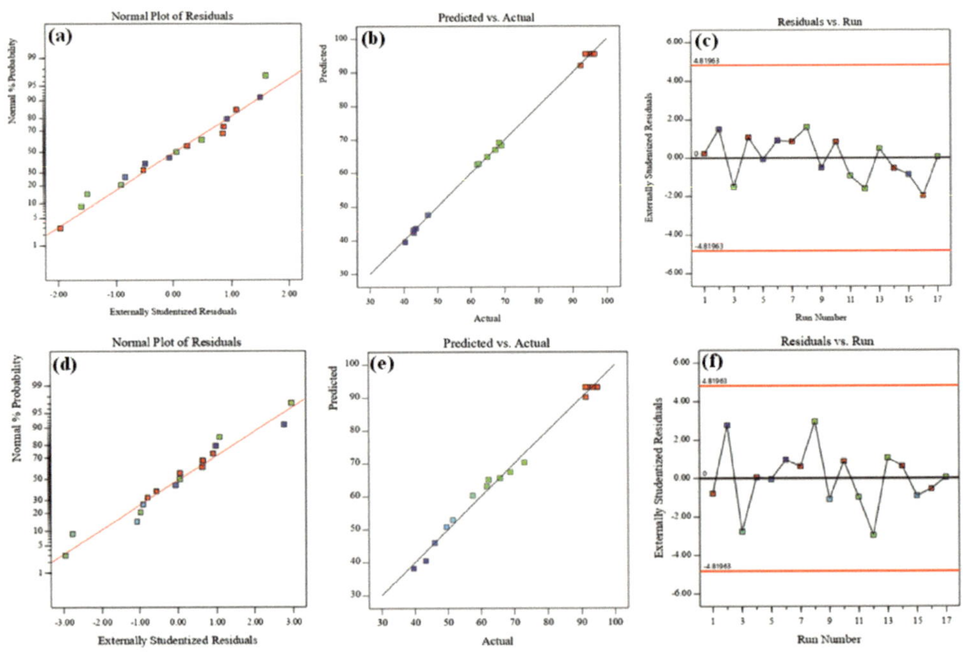

Figure 7.
Diagnostic plots for adsorption of PbII): Probability plot for the studentized residuals (a), comparison between actual and predicted values (b), plot of the externally studentized residuals vs. experimental run number (c), diagnostic plots for adsorption of Cu(II): Probability plot for the studentized residuals (d), comparison between actual and predicted values (e), plot of the externally studentized residuals vs. experimental run number (f).

proportion of normal probability in terms of residuals, it can be observed that the datum-points are approximately straight-line (**Figure 7a** and **d**). Into the other diagnostic plots, the actual responses were compared to their residuals based on predicted responses, suggesting that the quadratic model was required to predict removal efficiency in the experimental parameters (**Figure 7b** and **d**). In addition, as shown in the plot (**Figure 7c** and **f**) the data showed a good homogeneity. In **Figure** 7, the dispersion of the residuals is dispersed randomly about ±5, confirming that the results are coherent with the model. In the other diagnostic plots, actual responses were compared to their residues based on predicted responses, implying that the quadratic model was necessary to predict removal efficiency in experimental parameters.

By employing the RSM method, the evaluated models (Eqs. (2) and (3)) are used to design the 3-D graphs and find the optimal conditions for Pb (II) and Cu(II) removal efficiency. It can be seen from **Figures 8** and **9** that the retention of Pb(II) and Cu(II) ions onto BEC-ED increases with increases of the pH solution. The removal efficiency reached a maximum of around 6. When the pH is higher than 6 or lower than 5 the adsorption decreased rapidly. This could be explained by that in the acidic environment, the active groups responsible for the adsorption process exist mainly in the NH^{3+} form, and they prevent the retention of Pb (II) and Cu(I) ions on the amino groups of BEC. When the pH increases from 2 to 5, the active sites of the chelator become in the form of free NH_2 amines, which facilitate chelation on Pb(II) [60]. In addition, at high pH, the formation of lead and copper hydroxides (**Figure 10**) limits their adsorption on the BEC-ED surface, and as shown in **Figures 8** and **9**, at high pHs, the removal efficiency of Pb(II) and Cu(II) ions is significantly diminished. Contact time was also examined and as given in **Figures 8** and **9**. The results of the retention of Pb(II) and Cu(II) onto BEC-ED revealed that the maximum adsorption equilibrium can be achieved rapidly around 16 min.

In conclusion, based on the desirability optimization with three factors, the best removal efficiency was 99.52% and 97.5% for Pb(II) and Cu(II), respectively and was obtained at pH: 5.94, adsorbent amount: 22.2 mg, and contact time: 21.53 min (**Figure 11**).

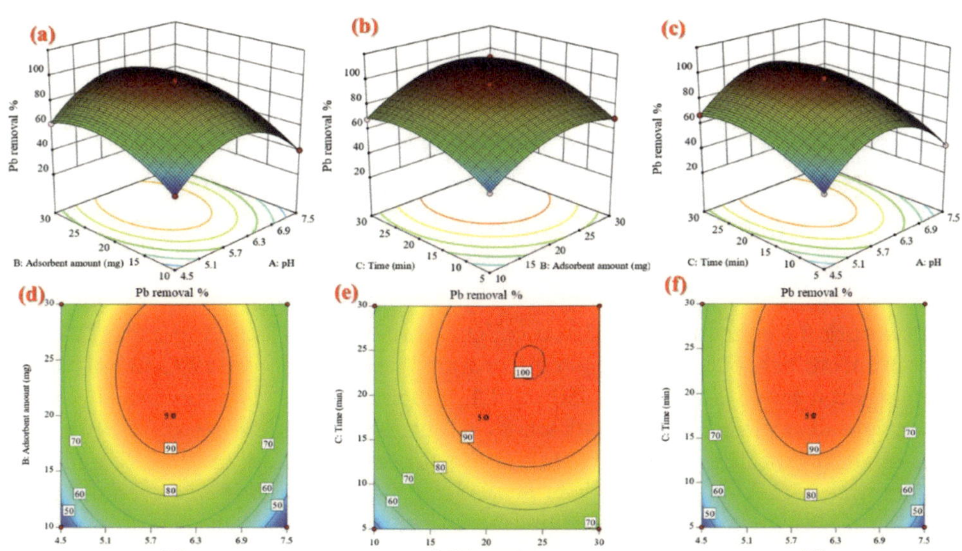

Figure 8.
(a, b and c) 3D response surface plot and, (d, e and f) contour plot for the effect of factors on the Pb(II) removal efficiency.

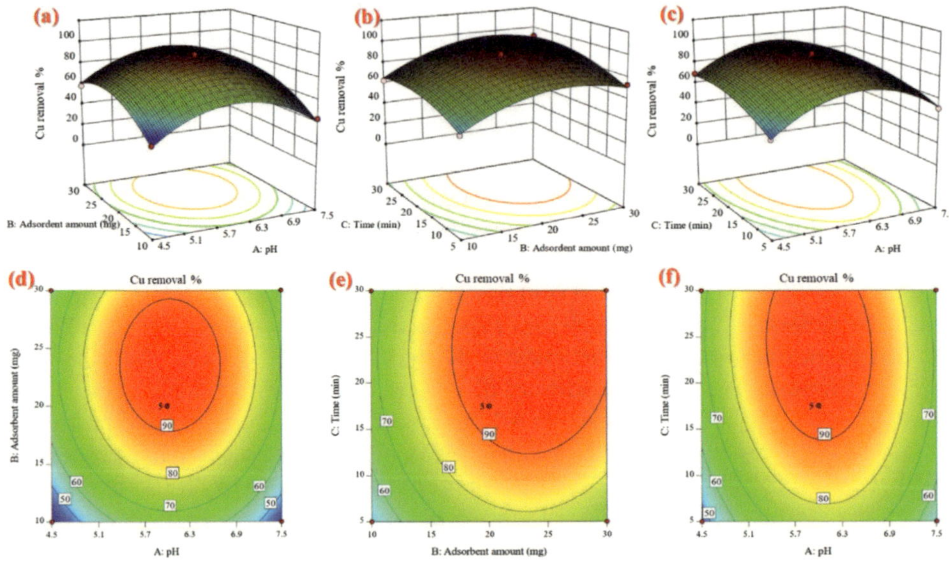

Figure 9.
(a, b, and c) 3D response surface plot and, (d, e and f) contour plot for the effect of factors on the Cu(II) removal efficiency.

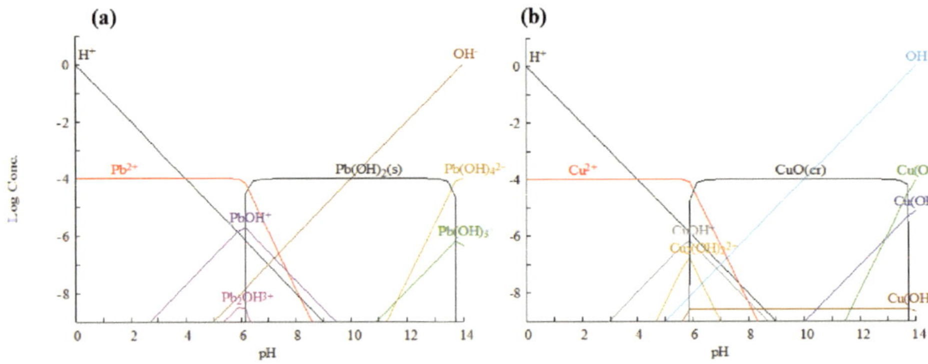

Figure 10.
Speciation diagram of lead (a) and copper (b) as a function of pH in ultrapure water, determined by the hydra/medusa program [81].

3.3 Equilibrium isotherms

Adsorption isotherms for lead and copper were made by carrying out batch adsorption studies. Lead and copper adsorption was studied onto BEC-ED in a large concentration range (from 15 to 250 mg/L), to better model the retention mechanisms. The adsorption experiments were performed at room temperature by using a mass of adsorbent 22.2 mg with 50 mL of the aqueous solution, at pH 5.94 and contact time 21.53 min. The quantity of the lead and copper ions adsorbed onto the BEC-ED at equilibrium, q_e (mg/g), and the adsorption percentage was calculated by the following Equations [82]:

$$q_e = \frac{(C_0 - C_e)}{m} \times V \tag{4}$$

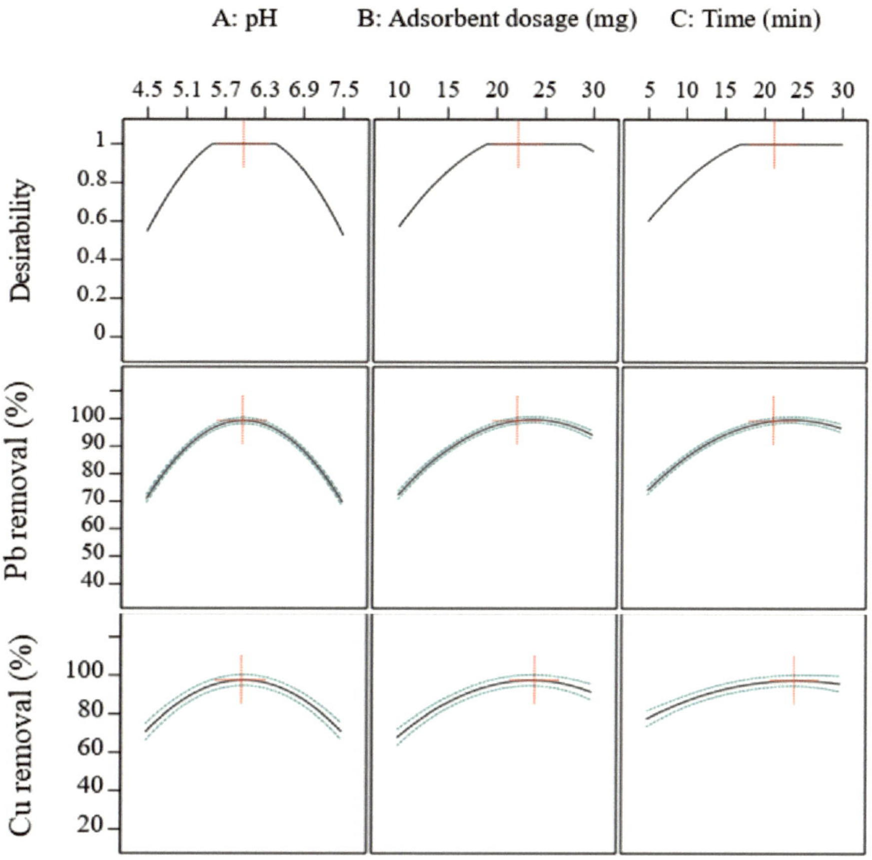

Figure 11.
Desirability approach function optimization for Pb(II) and Cu(II) in terms of removal efficiency (%) and desirability.

$$\%Adsorption = \left(1 - \frac{C_e}{C_0}\right) \times 100 \quad (5)$$

Where, C_0 and C_e are the metal ion initial concentration and concentration at equilibrium (mg/L), respectively. V is the volume of solution (L) and m is the adsorbent amount (g).

The adsorption isotherms of Cu(II) and Pb(II) on BEC-ED were modeled using the Freundlich (Eq. (6)) [83] and Langmuir (Eq. (7)) [84] models equations:

$$q_e = K_F C_e^{1/n} \quad (6)$$

Where, n and K_F are Freundlich constants represent the heterogeneity index, and the adsorption coefficient, respectively.

$$q_e = \frac{q_m K_L C_e}{1 + K_L q_m} \quad (7)$$

Where, K_L (L/mg) and q_m (mg/g) are the Langmuir constant and the maximum adsorption capacity, respectively.

The equilibrium isotherm obtained for lead and copper adsorption on BEC-ED is shown in **Figure 12**. Lead adsorption was greater than that of copper 43.85 mg/g. This could be explained by the higher reactivity of lead than copper, which can have

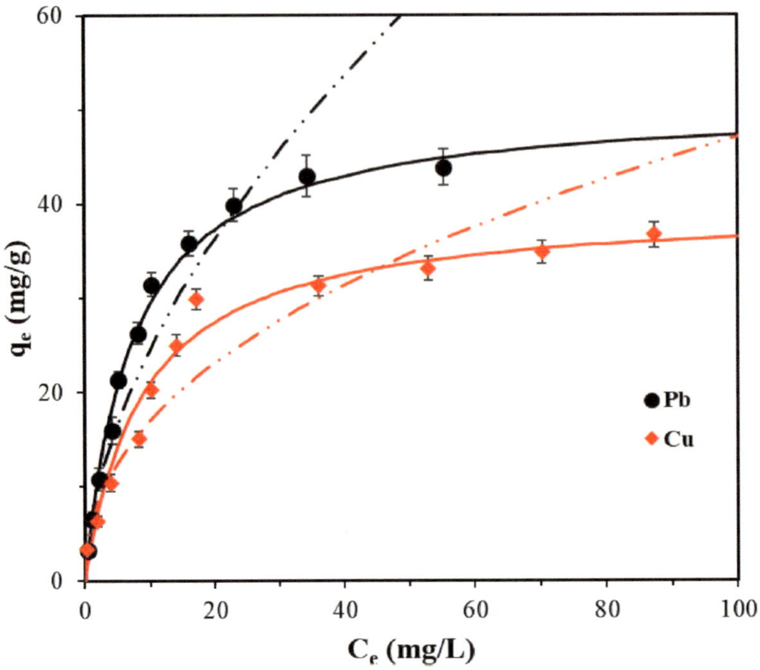

Figure 12.
Equilibrium isotherms for Pb and Cu adsorption on BEC-ED (solid lines and dash-dotted represent Langmuir and Freundlich fitting, respectively).

stronger interactions with the lone pairs of electrons of the nitrogen atoms of amino groups than that of the Cu, as previously demonstrated elsewhere [63, 85]. Therefore, the Pb(II) ions can rapidly form a stable complex with -NH_2 groups on the surface of the BEC- ED.

The adsorption isotherm was modeled both by Freundlich (Eq. (6)) and Langmuir (Eq. (7)) models. The results revealed that the equilibrium isotherms data (**Figure 12** and **Table 7**) correlated better with the Langmuir model with a maximum adsorption capacity estimated at 50.76 mg/g and 39.68 mg/g for Pb(II) and Cu (II), respectively. This implies that the BEC-ED surface is homogeneous, which indicates that lead and copper ions adsorption follows monolayer adsorption.

To better understand the retention mechanisms of copper and lead adsorption on BEC-ED, the effect of lead and copper ions adsorption on the morphology of BEC-ED was monitored by SEM analysis. **Figure 13** shows the possible interactions between BEC-ED and the Pb(II) and Cu (II) ions, as well as the proposed mechanism. SEM pictures showed distortion in the morphology of BEC-ED under the adsorption forces of Pb(II) and Cu(II) ions. However, this effect is probably caused by the interactions between the metal ions and the donor sites of the grafted groups (ED), where the internal compression of the laminated structure has caused a very remarkable separation of the polymeric layers. Indeed, the approximate calculation

	Langmuir				**Freundlich**		
	q_{exp} **(mg/g)**	q_m **(mg/g)**	K_L **(L/mg)**	R^2	**1/n**	K_F **(L/mg)**	R^2
Pb	43.85	50.76	0.137	0.996	0.443	6.1	0.951
Cu	36.64	39.68	0.112	0.990	0.5636	6.7	0.944

Table 7.
Freundlich and Langmuir constants for lead and copper adsorption on BEC-ED.

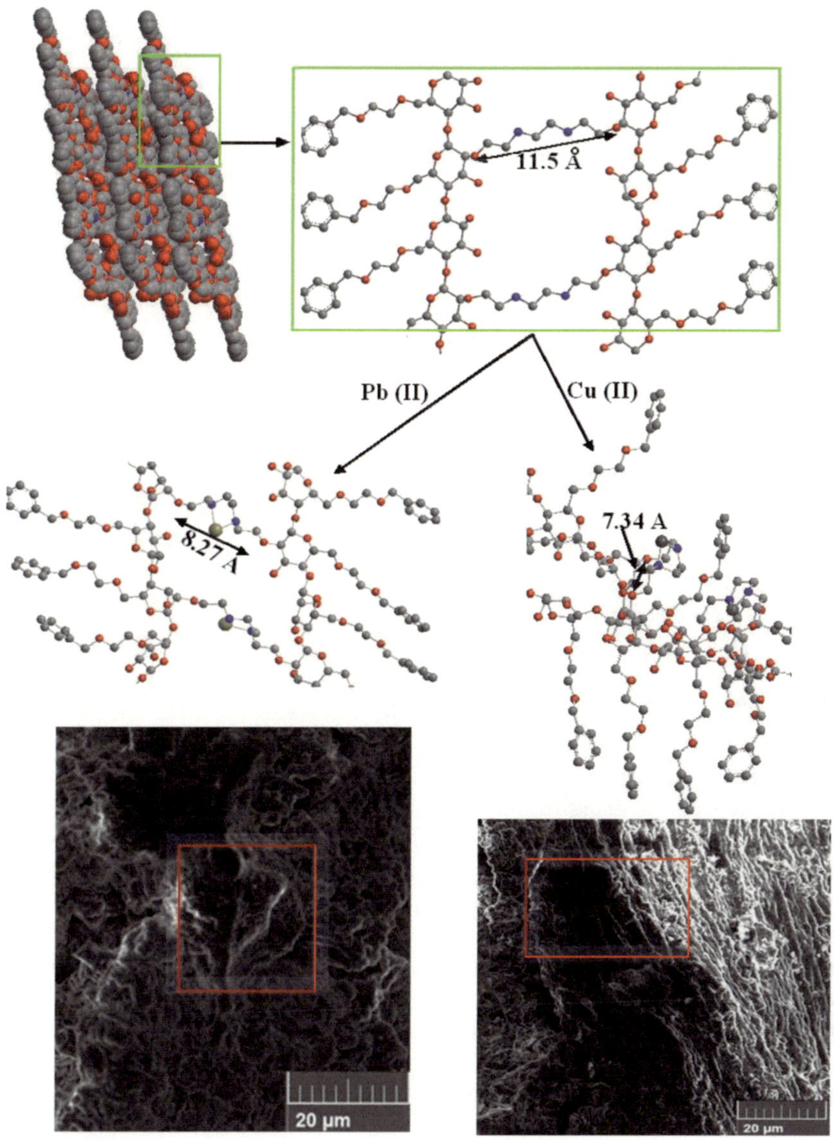

Figure 13.
Impact of the adsorption of lead and copper ions on the morphology of BEC-ED.

carried out to compare the inter-chain distances showed a decrease in the latter. The results obtained showed the capability of BEC-ED for adsorbing Cu(II) via metal interactions with NH_2 groups of ethylenediamine [56, 60], resulting in the decrease in the inter-layer distance from 11.50 Å to 07.34 Å for copper and 08.27 Å for lead.

For the purpose to assess the potential retention of lead and copper ions retention provided by BEC-ED compared to other adsorbents, the results achieved through this study were compared with the adsorption abilities of some conventional natural and synthetic cellulose in the literature (**Table 8**). It has been found that the lead and copper retention capacity of BEC-ED is among the higher results. Therefore, considering the retention capabilities of other adsorbents, accessibility, environment friendly biomaterial, and low cost, it may be concluded that the BEC-ED adsorbent demonstrated its ability to efficiently eliminate lead and copper ions in simple media.

Grafted cellulose adsorbent	Chelating group	Metal ions	Adsorption capacity (mg/g)	Ref.
Cellulose	Epichlorohydrin	Pb^{2+} Cu^{2+}	38.02 72.99	[86]
Microcrystalline cellulose	Tetrafluoroterephthalonitrile	Pb^{2+} Cu^{2+}	20.46 17.94	[87]
Cellulosic biopolymer	(alkali treatment)	Pb^{2+}, Cd^{2+} Zn^{2+}	67.24 44.42 16.85	[88]
CMC@ hydrogel	—	Cu^{2+}	2.30	[89]
BEC-ED	Ethylenediamine	Pb^{2+} Cu^{2+}	50.76 39.68	this work this work

Table 8.
Comparison of lead and copper adsorption capacity of BEC-ED with conventional natural and synthetic cellulose adsorbents.

4. Conclusion

A new green adsorbent, Benzyloxyethyl cellulose crosslinked ED (BEC-ED) was successfully synthesized. The proposal structures were confirmed using vibrational spectroscopy, X-ray diffraction patterns, SEM images, x-ray EDS spectra, and TGA thermograms. The results showed that the ED crosslinking reaction of BEC engendered new structural significant modification at the crystalline and morphological levels. Ethylenediamine crosslinked BEC has been used for the removal of lead (Pb) and copper (Cu) from an aqueous system. The results revealed that the equilibrium isotherms data correlated better with the Langmuir model with a maximum adsorption capacity estimated at 50.76 mg/g and 39.68 mg/g for Pb (II) and Cu (II), respectively. In addition, the results demonstrated that the capability of BEC-ED for adsorbing Cu (II) and Pb (II) was governed by metal–ligand interactions with NH_2 chelator sites of ethylenediamine, resulting in the decrease in the civility diameter from 11.50 Å to 07.34 Å for copper and 08.27 Å for lead. However, these distortions proved the lamellar structure, and the separation of the sheets was observed on the SEM images of BEC-ED after adsorption.

Acknowledgements

We thank greatly the anonymous reviewers for their careful review and valuable suggestions on the manuscript. The authors are thankful to the Head of Oujda's chemistry department, Prof. Abdelmonaem TALHAOUI, for providing all the facilities and subsidies necessary to carry out the research work of this article. Our warm thanks are addressed to Prof. Gharibi EL KHADIR, Director of the Solid Mineral Solid Chemistry Laboratory – Oujda, for his sincere and persistent contributions and devoted cooperation.

Conflict of interest

The authors declared that there is no conflict of interest.

Author details

Issam Jilal[1*], Soufian El Barkany[1*], Zahra Bahari[1], Youssef El Ouardi[2,3], Mohamed Loutou[1], Hassan Amhamdi[4], Mohamed Abou-Salama[1], Amin Salhi[4], Abderrahmane El Idrissi[5] and Katri Laatikainen[3]

1 Laboratory of Molecular Chemistry, Materials and Environment (LMCME), Department of Chemistry, Faculty Multidisciplinary Nador, Mohamed 1st University, Nador, Morocco

2 LIMOME Laboratory, Dhar El Mehraz Faculty of Sciences, Sidi Mohamed Ben Abdellah University, Fes, Morocco

3 Laboratory of Separation Technology, Lappeenranta University of Technology, Lappeenranta, Finland

4 Applied Chemistry Unit, Sciences and Technologies Faculty, Abdelmalek Essaadi University, Al Hoceima, Morocco

5 Laboratory Applied Chemistry and Environmental (LCAE-URAC18), Faculty of Sciences of Oujda, Mohamed1stUniversity, Oujda, Morocco

*Address all correspondence to: issamjilal@gmail.com
and el.barkany011@gmail.com

References

[1] Hokkanen S, Bhatnagar A, Sillanpää M. A review on modification methods to cellulose-based adsorbents to improve adsorption capacity. Water Research. 2016;**91**:156-173. DOI: 10.1016/j.watres.2016.01.008

[2] Pandiyan J, Mahboob S, Govindarajan M, Al-ghanim KA, Ahmed Z, Al-mulhm N, et al. An assessment of level of heavy metals pollution in the water, sediment and aquatic organisms: A perspective of tackling environmental threats for food security. Saudi Journal of Biological Sciences. 2021;**28**(2):1218-1225. DOI: 10.1016/j.sjbs.2020.11.072

[3] Karimi A, Haghnia GH, Ayoubi S, Safari T. Impacts of geology and land use on magnetic susceptibility and selected heavy metals in surface soils of Mashhad plain, northeastern Iran. Journal of Applied Geophysics. 2017; **138**:127-134. DOI: 10.1016/j.jappgeo.2017.01.022

[4] Kim HS, Kim YJ, Seo YR. An overview of carcinogenic heavy metal: molecular toxicity mechanism and prevention. Journal of cancer Prevention. 2015;**20**(4):232. DOI: 10.15430/JCP.2015.20.4.232

[5] Peng S, Dai M, Zhang J, Zhang M, Shi Q, Liang B, Zheng T. Dynamics of ecological risks associated with heavy metals in sediments during the construction process of the Yangtze River deepwater channel. Journal of Cleaner Production. 2020;269:122231. DOI: 10.1016/j.jclepro.2020.122231

[6] Zhang J, Shi Q, Fan S, Zhang Y, Zhang M, Zhang J. Distinction between Cr and other heavy–metal–resistant bacteria involved in C/N cycling in contaminated soils of copper producing sites. Journal of Hazardous Materials. 2021;402:123454. DOI: 10.1016/j.jhazmat.2020.123454

[7] Nriagu JO, Pacyna JM. Quantitative assessment of worldwide contamination of air, water and soils by trace metals. Nature. 1988;**333**(6169):134-139. DOI: 10.1038/333134a0

[8] Ju X-J, Zhang S-B, Zhou M-Y, Xie R, Yang L, Chu L-Y. Novel heavy-metal adsorption material: ion-recognition P (NIPAM-co-BCAm) hydrogels for removal of lead (II) ions. Journal of Hazardous Materials. 2009;**167**(1-3): 114-118. DOI: 10.1016/j.jhazmat.2008.12.089

[9] Priastomo Y, Setiawan HR, Kurniawan YS, Ohto K. Simultaneous removal of lead (II), chromium (III), and copper (II) heavy metal ions through an adsorption process using C-phenylcalix [4] pyrogallolarene material. Journal of Environmental Chemical Engineering. 2020;8(4): 103971. DOI: 10.1016/j.jece.2020.103971

[10] Bankole MT, Abdulkareem AS, Mohammed IA, Ochigbo SS, Tijani JO, Abubakre OK, et al. Selected heavy metals removal from electroplating wastewater by purified and polyhydroxylbutyrate functionalized carbon nanotubes adsorbents. Scientific Repots. 2019;**9**(1):1-19. DOI: 10.1038/s41598-018-37899-4

[11] Yatribi A, Nejmeddine A, Boukhars L. Contribution à la valorisation industrielle des déchets de tanneries: Cas de la cimenterie. In: Annales de Chimie Science Des Materiaux. Vol 26. Elsevier; 2001: 107-112.

[12] Kushwaha A, Hans N, Kumar S, Rani R. A critical review on speciation, mobilization and toxicity of lead in soil-microbe-plant system and bioremediation strategies. Ecotoxicology Environmental Safety. 2018;**147**:1035-1045. DOI: 10.1016/j.ecoenv.2017.09.049

[13] Mohan S, Sreelakshmi G. Fixed bed column study for heavy metal removal using phosphate treated rice husk. Journal of Hazardous Materials. 2008; **153**(1-2):75-82. DOI: 10.1016/j.jhazmat.2007.08.021

[14] Zhang Y. 100 years of Pb deposition and transport in soils in Champaign, Illinois. USA. Water Air and Soil Pollution. 2003;**146**(1):197-210. DOI: 10..1023/A:1023957226204

[15] Badawy NA, El-Bayaa AA, Abdel-Aal AY, Garamon SE. Chromatographic separations and recovery of lead ions from a synthetic binary mixtures of some heavy metal using cation exchange resin. Journal of Hazardous Materials. 2009;**166**(2-3):1266-1271. DOI: 10.1016/j.jhazmat.2008.12.044

[16] Nordberg GF, Fowler BA. Nordberg M. Handbook on the Toxicology of Metals: Academic press; 2014

[17] Jiang X, Liu W, Xu H, Cui X, Li J, Chen J, Zheng B. Characterizations of heavy metal contamination, microbial community, and resistance genes in a tailing of the largest copper mine in China. Environmental Pollution. Published online 2021:116947. DOI: 10.1016/j.envpol.2021.116947

[18] Li L, Pan D, Li B, Wu Y, Wang H, Gu Y, et al. Patterns and challenges in the copper industry in China. Resources, Conservation and Recycling. 2017;**127**: 1-7. DOI: 10.1016/j.resconrec.2017.07.046

[19] Chen Q, Yao Y, Li X, Lu J, Zhou J, Huang Z. Comparison of heavy metal removals from aqueous solutions by chemical precipitation and characteristics of precipitates. Journal of water Process Engineering. 2018;**26**: 289-300. DOI: 10.1016/j.jwpe.2018.11.003

[20] Zhang Y, Duan X. Chemical precipitation of heavy metals from wastewater by using the synthetical magnesium hydroxy carbonate. Water Science and Technology. 2020;**81**(6): 1130-1136. DOI: 10.2166/wst.2020.208

[21] Hube S, Eskafi M, Hrafnkelsdóttir KF, Bjarnadóttir B, Bjarnadóttir M, Axelsdóttir S, Wu B. Direct membrane filtration for wastewater treatment and resource recovery: A review. Science of the Total Environment. 2020;710:136375. DOI: 10.1016/j.scitotenv.2019.136375

[22] Erçarıkcı E, Alanyalıoğlu M. Dual-Functional Graphene-Based Flexible Material for Membrane Filtration and Electrochemical Sensing of Heavy Metal Ions. IEEE Sensors Journal. 2020;**21**(3): 2468-2475. DOI: 10.1109/JSEN.2020.3021988

[23] Sun Y, Zhou S, Pan S-Y, Zhu S, Yu Y, Zheng H. Performance evaluation and optimization of flocculation process for removing heavy metal. Chemical Engineering Journal. 2020;385:123911. DOI: 10.1016/j.cej.2019.123911

[24] Swain K, Abbassi B, Kinsley C. Combined electrocoagulation and chemical coagulation in treating brewery wastewater. Water. 2020; **12**(3):726. DOI: 10.3390/w12030726

[25] Yeh G, Hoang H-G, Lin C, Bui X, Tran H, Shern C, et al. Assessment of heavy metal contamination and adverse biological effects of an industrially affected river. Environmental Science and Pollution Research. 2020;**27**(28): 34770-34780. DOI: 10.1007/s11356-020-07737-0

[26] Ibrahim Y, Naddeo V, Banat F, Hasan SW. Preparation of novel polyvinylidene fluoride (PVDF)-Tin (IV) oxide (SnO2) ion exchange mixed matrix membranes for the removal of heavy metals from aqueous solutions. Separation and Purification Technology. 2020;250:117250. DOI: 10.1016/j.seppur.2020.117250

[27] Bożęcka A, Orlof-Naturalna M, Sanak-Rydlewska S. Removal of lead, cadmium and copper ions from aqueous solutions by using ion exchange resin C 160. Gospod Surowcami Miner. 2016;**32**. DOI: 10.1515/gospo-2016-0033

[28] Dai K, Chen H, Peng T, Ke D, Yi H. Photocatalytic degradation of methyl orange in aqueous suspension of mesoporous titania nanoparticles. Chemosphere. 2007;**69**(9):1361-1367. DOI: 10.1016/j.chemosphere. 2007.05.021

[29] Zhao Y, Kang S, Qin L, Wang W, Zhang T, Song S, Komarneni S. Self-assembled gels of Fe-chitosan/ montmorillonite nanosheets: dye degradation by the synergistic effect of adsorption and photo-Fenton reaction. Chemical Engineering Journal. 2020; 379:122322. DOI: 10.1016/j. cej.2019.122322

[30] Saravanan R, Gupta VK, Narayanan V, Stephen A. Comparative study on photocatalytic activity of ZnO prepared by different methods. Journal of Molecular Liquids. 2013;**181**:133-141. DOI: 10.1016/j.molliq.2013.02.023Get

[31] Saravanan R, Khan MM, Gupta VK, Mosquera E, Gracia F, Narayanan V, et al. ZnO/Ag/Mn 2 O 3 nanocomposite for visible light-induced industrial textile effluent degradation, uric acid and ascorbic acid sensing and antimicrobial activity. RSC Advences. 2015;**5**(44):34645-34651. DOI: 10.1039/ C5RA02557E

[32] Qi Y, Zhu L, Shen X, Sotto A, Gao C, Shen J. Polythyleneimine-modified original positive charged nanofiltration membrane: Removal of heavy metal ions and dyes. Separation and Purification Technology. 2019;**222**: 117-124. DOI: 10.1016/j. seppur.2019.03.083

[33] Jilal I, El Barkany S, Bahari Z, Sundman O, El Idrissi A, Abou-Salama M, et al. New quaternized cellulose based on hydroxyethyl cellulose (HEC) grafted EDTA: synthesis, characterization and application for Pb (II) and Cu (II) removal. Carbohydrate Polymers. 2018;**180**:156-167. DOI: 10.1016/j.carbpol.2017.10.012

[34] Liu Q, Li Y, Chen H, Lu J, Yu G, Moslang M, Zhou Y. Superior adsorption capacity of functionalised straw adsorbent for dyes and heavy-metal ions. Journal of Hazardous Materials. 2020;382:121040. DOI: 10.1016/j.jhazmat.2019.121040

[35] Choi HY, Bae JH, Hasegawa Y,An S, Kim I, Lee H, KimM. Thiol-functionalized cellulose nanofiber membranes for the effective adsorption of heavy metal ions in water. Carbohydrate Polymers. 2020;234: 115881. DOI: 10.1016/j. carbpol.2020.115881

[36] Jiang C, Wang X, Hou B, Hao C, Li X, Wu J. Construction of a lignosulfonate–lysine hydrogel for the adsorption of heavy metal ions. Journal Agricultural and Food Chemistry. 2020; **68**(10):3050-3060. DOI: 10.1021/acs. jafc.9b07540

[37] Niu M, Li G, Cao L, Wang X, Wang W. Preparation of sulphate aluminate cement amended bentonite and its use in heavy metal adsorption. Journal of Cleaner Production. 2020;256: 120700. DOI: 10.1016/j. jclepro.2020.120700

[38] Yap PL, Auyoong YL, Hassan K, Farivar F, Tran D, Ma J, Losic D. Multithiol functionalized graphene bio-sponge via photoinitiated thiol-ene click chemistry for efficient heavy metal ions adsorption. Chemical Engineering Journal. 2020;395:124965. DOI: 10.1016/ j.cej.2020.124965

[39] Lacuesta AC, Herrera MU, Manalo R, Balela MDL. Fabrication of kapok paper-zinc oxide-polyaniline

hybrid nanocomposite for methyl orange removal. Surface and Coatings Technology. 2018;**350**:971-976. DOI: 10.1016/j.surfcoat.2018.03.043

[40] Herrera MU, Futalan CM, Gapusan R, Balela MDL. Removal of methyl orange dye and copper (II) ions from aqueous solution using polyaniline-coated kapok (Ceiba pentandra) fibers. Water Science and Technology. 2018;**78**(5):1137-1147. DOI: 10.2166/wst.2018.385

[41] Balela MDL, Intila NM, Salvanera SR. Adsorptive Removal of Lead Ions in Aqueous Solution by Kapok-Polyacrylonitrile Nanocomposites. Materials Today Proceeding. 2019;**17**:672-678. DOI: 10.1016/j.matpr.2019.06.349

[42] Awual MR, Yaita T, Okamoto Y. A novel ligand based dual conjugate adsorbent for cobalt (II) and copper (II) ions capturing from water. Sensors and Actuators B Chemical. 2014;**203**:71-80. DOI: 10.1016/j.snb.2014.06.088

[43] El Ouardi Y, Branger C, Toufik H, Laatikainen K, Ouammou A, Lenoble V. An insight of enhanced natural material (calcined diatomite) efficiency in nickel and silver retention: Application to natural effluents. Environmental Technology and Innovation. 2020;18: 100768. DOI: 10.1016/j.eti.2020.100768

[44] El-Saied H, Mostafa AM, Hasanin MS, Mwafy EA, Mohammed AA. Synthesis of antimicrobial cellulosic derivative and its catalytic activity. Journal of King Saud University-Science. 2020;**32**(1): 436-442. DOI: 10.1016/j. jksus.2018.06.007

[45] Abou-Yousef H, Dacrory S, Hasanin M, Saber E, Kamel S. Biocompatible hydrogel based on aldehyde-functionalized cellulose and chitosan for potential control drug release. Sustainable Chemistry and Pharmacy. 2021;21:100419. DOI: 10.1016/j.scp.2021.100419

[46] Salama A. Cellulose/silk fibroin assisted calcium phosphate growth: Novel biocomposite for dye adsorption. International Journal of Biological Macromolecules. 2020;**165**:1970-1977. DOI: 10.1016/j.ijbiomac.2020.10.074

[47] Salama A. New sustainable hybrid material as adsorbent for dye removal from aqueous solutions. Journal of Colloid and Interface Science. 2017;**487**: 348-353. DOI: 10.1016/j.jcis.2016.10.034

[48] Dacrory S, Hashem AH, Hasanin M. Antifungal activity and hemocompatibility of cellulose based amino acid functionalized nano-biocomplex. Environmental Nanotechnology, Monitoring & Management. Published online 2021: 100453. DOI: 10.1016/j. enmm.2021.100453

[49] Putro JN, Santoso SP, Ismadji S, Ju Y-H. Investigation of heavy metal adsorption in binary system by nanocrystalline cellulose–bentonite nanocomposite: improvement on extended Langmuir isotherm model. Microporous and Mesoporous Materials. 2017;**246**:166-177. DOI: 10.1016/j. micromeso.2017.03.032

[50] Chen Q, Zheng J, Wen L, Yang C, Zhang L. A multi-functional-group modified cellulose for enhanced heavy metal cadmium adsorption: Performance and quantum chemical mechanism. Chemosphere. 2019;**224**: 509-518. DOI: 10.1016/j. chemosphere.2019.02.138

[51] Godiya CB, Cheng X, Li D, Chen Z, Lu X. Carboxymethyl cellulose/ polyacrylamide composite hydrogel for cascaded treatment/reuse of heavy metal ions in wastewater. Journal of Hazardous Materials. 2019;**364**:28-38. DOI: 10.1016/j.jhazmat.2018.09.076

[52] Dacrory S, Abou-Yousef H, Kamel S, Turky G. Development of biodegradable semiconducting foam based on micro-fibrillated cellulose/Cu-NPs. International Journal of Biological Macromolecules. 2019;**132**:351-359. DOI: 10.1016/j.ijbiomac.2019.03.156

[53] El Barkany S, El Idrissi A, Zanagui C, Jilal I, Tabghat F, Abou-Salama M, et al. New branched chain cellulose derivatives based on esparto Stipa tenacissima of eastern Morocco: synthesis and characterization. Journal of Materials Environment Science. 2017; **8**(4):1195-1210

[54] Jilal I, El Barkany S, Bahari Z, Sundman O, El Idrissi A, Salhi A, et al. Unconventional synthesis, characterization and theoretical study (HF and DFT computations) of new cellulosic copper complex: benzyloxyethyl cellulose copper (CuBEC). Cellulose. 2018;**25**(8): 4375-4388. DOI: 10.1007/s10570-018-1909-x

[55] Chaouf S, El Barkany S, Amhamdi H, Jilal I, El Ouardi Y, Abou-Salama M, et al. Low degree of substitution of cellulose acrylate based green polyelectrolyte: Synthesis, characterization and application to the removal of Cu (II) ions and colloidal Fe (OH) 3 turbidity. Materials Today Proceeding. 2020;**31**:S175-S182. DOI: 10.1016/j.matpr.2020.07.512

[56] Yakout AA, El-Sokkary RH, Shreadah MA, Hamid OGA. Cross-linked graphene oxide sheets via modified extracted cellulose with high metal adsorption. Carbohydrate Polymers. 2017;**172**:20-27. DOI: 10.1016/j.carbpol.2017.05.004

[57] Guo Z, Fan J, Zhang J, Kang Y, Liu H, Jiang L, et al. Sorption heavy metal ions by activated carbons with well-developed microporosity and amino groups derived from Phragmites australis by ammonium phosphates activation. Journal of the Taiwan Institute of Chemical Engineers. 2016; **58**:290-296. DOI: 10.1016/j.jtice.2015.05.041

[58] Wu F-C, Tseng R-L, Juang R-S. Kinetic modeling of liquid-phase adsorption of reactive dyes and metal ions on chitosan. Water Research. 2001; **35**(3):613-618. DOI: 10.1016/S0043-1354(00)00307-9

[59] Yoshitake H, Yokoi T, Tatsumi T. Adsorption behavior of arsenate at transition metal cations captured by amino-functionalized mesoporous silicas. Chemistry of Materials. 2003; **15**(8):1713-1721. DOI: 10.1021/cm0218007

[60] Racho P, Phalathip P. Modified nylon fibers with amino chelating groups for heavy metal removal. Energy Procedia. 2017;**118**:195-200. DOI: 10.1016/j.egypro.2017.07.026

[61] Liang X, Liang B, Wei J, Zhong S, Zhang R, Yin Y, et al. A cellulose-based adsorbent with pendant groups of quaternary ammonium and amino for enhanced capture of aqueous Cr (VI). International Journal of Biological Macromolecules. 2020;**148**:802-810. DOI: 10.1016/j.ijbiomac.2020.01.184

[62] Wang J, Zhao L, Duan W, Han L, Chen Y. Adsorption of aqueous Cr (VI) by novel fibrous adsorbent with amino and quaternary ammonium groups. Industrial & Engineering Chemistry Research. 2012;**51**(42):13655-13662. DOI: 10.1021/ie3013874

[63] Wang K, Gu J, Yin N. Efficient removal of Pb (II) and Cd (II) using NH2-functionalized Zr-MOFs via rapid microwave-promoted synthesis. Industrial & Engineering Chemistry Research. 2017;**56**(7):1880-1887. DOI: 10.1021/acs.iecr.6b04997

[64] Edebali S, Pehlivan E. Evaluation of chelate and cation exchange resins to

remove copper ions. Powder Technology. 2016;**301**:520-525. DOI: 10.1016/j.powtec.2016.06.011

[65] Gupta S, Varshney PK. Effect of plasticizer concentration on structural and electrical properties of hydroxyethyl cellulose (HEC)-based polymer electrolyte. Ionics (Kiel). 2017; **23**(6):1613-1617. DOI: 10.1007/s11581-017-2116-8

[66] Fiayaz M, Zia KM, Javaid MA, Rehman S, Chatha SAS, Zuber M. Synthesis and characterization of hydroxyethyl cellulose (HEC)-TiO 2-based polyurethane bionanocomposites. Korean Journal of Chemical Enginnering. 2020;**37**(12):2351-2358. DOI: 10.1007/s11814-020-0664-5

[67] El Idrissi A, El Barkany S, Amhamdi H, Maaroufi A. Synthesis and characterization of the new cellulose derivative films based on the hydroxyethyl cellulose prepared from esparto "stipa tenacissima" cellulose of Eastern Morocco. II. Esterification with acyl chlorides in a homogeneous medium. Journal of Applied Polymer Science. 2013;**127**(5):3633-3644. DOI: 10.1002/app.37982

[68] Nizan NSNH, Zulkifli FH. Reinforcement of hydroxyethyl cellulose/poly (vinyl alcohol) with cellulose nanocrystal as a bone tissue engineering scaffold. Journal of Polymer Research. 2020;**27**:1-9. DOI: 10.1007/s10965-020-02112-6

[69] Das A, Kundu S, Ghosh SK, Basu A, Gupta M, Mukherjee A. Guar gum cinnamate ouzo nanoparticles for bacterial contact killing in water environment. Carbohydrate Research. 2020;491:107983. DOI: 10.1016/j.carres.2020.107983

[70] Jafari M, Davachi SM, Mohammadi-Rovshandeh J, Pouresmaeel-Selakjani P. Preparation and characterization of bionanocomposites based on benzylated wheat straw and nanoclay. Journal of Polymers Environmental. 2018;**26**(3): 913-925. DOI: 10.1007/s10924-017-0997-2

[71] Kundu S, Das A, Basu A, Abdullah MF, Mukherjee A. Guar gum benzoate nanoparticle reinforced gelatin films for enhanced thermal insulation, mechanical and antimicrobial properties. Carbohydrate Polymers. 2017;**170**:89-98. DOI: 10.1016/j.carbpol.2017.04.056

[72] Li M-F, Sun S-N, Xu F, Sun R-C. Cold NaOH/urea aqueous dissolved cellulose for benzylation: Synthesis and characterization. Europian Polymer Journal. 2011;**47**(9):1817-1826. DOI: 10.1016/j.eurpolymj.2011.06.013

[73] da Silva Filho EC, de Melo JCP, Airoldi C. Preparation of ethylenediamine-anchored cellulose and determination of thermochemical data for the interaction between cations and basic centers at the solid/liquid interface. Carbohydrate Research. 2006; **341**(17):2842-2850. DOI: 10.1016/j.carres.2006.09.004

[74] de Melo JCP, da Silva Filho EC, Santana SAA, Airoldi C. Maleic anhydride incorporated onto cellulose and thermodynamics of cation-exchange process at the solid/liquid interface. Colloids Surfaces A: Physicochemical Engineering Aspects. 2009;**346**(1-3):138-145. DOI: 10.1016/j.colsurfa.2009.06.006

[75] Pavia DL, Lampman GM, Kriz GS, Vyvyan JA. Introduction to Spectroscopy. Nelson Education. 2014

[76] Da Silva Filho EC, de Barros Júnior JF, Santana SAA, de Melo JCP, Airoldi C. Thermodynamics of cation/basic center interactions from ethylene-1, 2-diamine? pentane-2, 4-dione cellulose incorporated. Glob. J Phys Chem. 2011;**2**:277-286

[77] El Ghali A, Baouab MHV, Roudesli MS. Preparation, characterization and application of a [copper (II)/ethylenediamine–cotton] complex for the removal of AB25 from aqueous solution in a laboratory scale column. Chemical Engineering Journal. 2011;**174**(1):18-26. DOI: 10.1016/j.cej.2011.07.046

[78] Dhahri A, Beyou E, Alcouffe P, Serghei A, Baouab MHV. Synthesis and characterization of Au-immobilized nanoparticles onto cellulose-ethylenediamine-grafted reduced graphite oxide sheets. Materials Chemistry and Physics. 2016;**171**: 303-311. DOI: 10.1016/j.matchemphys.2016.01.021

[79] Ramírez JAÁ, Hoyos CG, Arroyo S, Cerrutti P, Foresti ML. Acetylation of bacterial cellulose catalyzed by citric acid: Use of reaction conditions for tailoring the esterification extent. Carbohydrate Polymers. 2016;**153**: 686-695. DOI: 10.1016/j.carbpol.2016.08.009

[80] Jung KW, Kim DH, Kim HW, Shin HS. Optimization of combined (acid + thermal) pretreatment for fermentative hydrogen production from Laminaria japonica using response surface methodology (RSM). International Journal of Hydrogen Energy. 2011;**36**(16):9626-9631. DOI: 10.1016/j.ijhydene.2011.05.050

[81] Ignasi Puigdomenech. Hydra/Medusa Chemical Equilibrium Database and Plotting Software (2015).

[82] El Ouardi Y, Branger C, Laatikainen K, Durrieu G, Mounier S, Ouammou A, et al. Impact of thermal treatment on bentonite retention ability toward nickel and silver retention. Separation Science and Technology. 2020;**55**:18):1-18)11. DOI: 10.1080/01496395.2020.1839772

[83] Freundlich H. Über die Adsorption in Lösungen. Zeitschrift für Physikalische Chemie. 1906;**57**(4): 385-470

[84] Langmuir I. The adsorption of gases on plane surfaces of glass, mica and platinum. Journal of the American Chemcal Society. 1918;**345**(1914): 1361-1403. DOI: 10.1021/ja02242a004

[85] Ahmadijokani F, Tajahmadi S, Bahi A, Molavi H, Rezakaezmi M, Ko F, et al. Ethylenediamine-functionalized Zr-based MOF for efficient removal of heavy metal ions from water. Chemosphere. 2021;**264**. DOI: 10.1016/j.chemosphere.2020.128466

[86] Taksitta K, Sujarit P, Ratanawimarnwong N, Donpudsa S, Songsrirote K. Development of tannin-immobilized cellulose fiber extracted from coconut husk and the application as a biosorbent to remove heavy metal ions. Environmental Nanotechnology, Monitoring & Management. 2020;14 (October):100389. DOI: 10.1016/j.enmm.2020.100389

[87] Cao J, Fei D, Tian X, Zhu Y, Wang S, Zhang Y, et al. Novel modified microcrystalline cellulose-based porous material for fast and effective heavy-metal removal from aqueous solution. Cellulose. 2017;**24**(12):5565-5577. DOI: 10.1007/s10570-017-1504-6

[88] Singha AS, Guleria A. Use of Low Cost Cellulosic Biopolymer Based Adsorbent for the Removal of Toxic Metal Ions from the Aqueous Solution. Separation Science and Technology. 2014;49(16):2557-2567. DOI: 10.1080/01496395.2014.929146

[89] Baiya C, Nannuan L, Tassanapukdee Y, Chailapakul O, Songsrirote K. The Synthesis of Carboxymethyl Cellulose-Based Hydrogel from Sugarcane Bagasse Using Microwave-Assisted Irradiation for Selective Adsorption of Copper(II) Ions. Environmental Progress & Sustainable Energy. 2019;38(s1):S157-S165. DOI: 10.1002/ep.12950

Chapter 8

Conversion of Cellulose into Value-Added Products

Lethiwe D. Mthembu, Rishi Gupta and Nirmala Deenadayalu

Abstract

Rapid industrialization has led to development of various platform chemicals and fossil fuel refinery is one of the mainstreams for their production. However continuous depletion of fossil fuels reserves has led an urge to look for alternate source of feedstocks. Among various renewable sources, biomass is found to be most sustainable as it is replaced naturally. Biomass by virtue of its nature is comprised of various recalcitrant polymers and cellulose is one of them, which can be used for the generation of various platform chemicals. This chapter gives a background of cellulose and illustrate platform chemicals that can be produced from cellulose.

Keywords: cellulose, renewable resources, biomass, glucose, value-added products

1. Introduction

Till date most platform chemicals were produced from petroleum refinery where crude oil is converted into fuels and chemicals that provide multiple products and revenue streams but depletion of fossil fuels and global warming are the two major concerns in chemical manufacturing. These concerns have led to the utilization of renewable resources such as biomass as an alternative feedstock for the development of Biorefinery [1]. Biorefineries are facilities that convert renewable biomass into biofuels, chemicals, and materials such as plastics and polymers [2]. The products developed in biorefinery are usually called bioproducts because it is based on the sustainable manufacture of products from biomass, this encompasses all processes, from feedstock production and supply to processing stages and end product manufacturing, as well as research, development, and commercialization procedures. The advantage of renewable resources over fossil fuels is that renewable resources can be replaced over time by natural processes. Moreover, complying green chemistry principle these renewable resources-based methods are also environmental benign. The global market for biorefinery products, by application includes industrial, manufacturing, transportation, flame retardants, safe food and supply, environment, communication, construction/housing, recreation, agriculture, health and hygiene, and energy. As an estimate, the worldwide market for biorefinery products would increase from $586.8 billion in 2020 to $867.7 billion by 2025, with a compound annual growth rate (CAGR) of 8.1 percent from 2020 to 2025 [3].

The main raw material for a biorefinery is the plant biomass, whose structural carbohydrate polymers (cellulose and hemicellulose) can be used to develop a

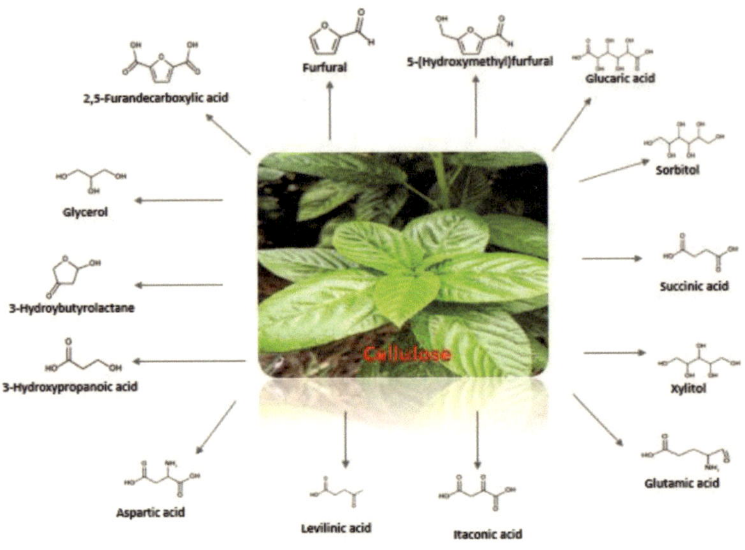

Figure 1.
Platform chemicals that are produced from biomass (cellulose).

variety of platform chemicals. The Department of Energy (DOE) of the United States (US) discovered 12 platform chemicals that can be produced from cellulose in biomass: 1,4-diacids (succinic, fumaric, and malic), 2,5-furan dicarboxylic acid, 3-hydroxypropionic acid, aspartic acid, glucaric acid, glutamic acid, itaconic acid, levulinic acid, hydroxybutyrolactone, glycerol, sorbitol, and xylitol/arabinitol [4]. A brief outline of the platform chemicals that are produced from biomass (cellulose) is illustrated in **Figure 1**.

The application of platform chemicals which makes them very interesting for industrial purpose. In the following section the focus will be on the cellulose, various pre-treatment used in biomass, more information about the platform chemicals and how they are produced from biomass, examples of platform chemicals produced from biomass.

2. Cellulose

Cellulose $(C_6H_{10}O_5)n$ is a stable crystalline homopolysaccharide polymer composed of the anhydroglucopyranose monomers formed via β-1,4-D-glucose linkage of carbohydroglucose units and contains several intra and intramolecular hydrogen bonds as well as intermolecular van der Waals forces [5]. Cellulose is insoluble in water and unlikely to be dissolve in most organic solvents due to the strong intramolecular and intermolecular hydrogen bonds which increases the rigidity [6]. Natural cellulose molecules are bundle of microfibrils (i.e, crystalline and amorphous regions) that are stabilized by hydrogen bonds between hydroxyl groups [7]. **Figure 2** shows the structure of one chain of the polymer cellulose [8].

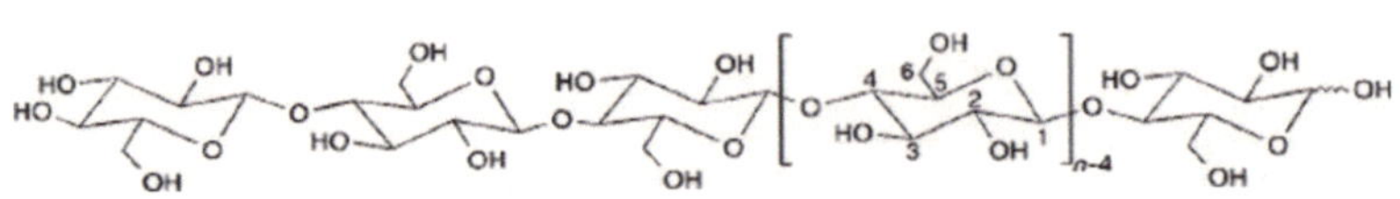

Figure 2.
Chemical structure of cellulose [8].

3. Pre-treatment of plant biomass

Cellulose, the most abundant natural polymer, is predominantly found in plant cell wall. However, besides cellulose, plant cell wall also comprised of hemicellulose and lignin, which make an interwoven structure in plant cell wall to make it recalcitrant for microbial and/or enzymatic attack [9]. **Table 1** illustrate how the compositional analysis of different type of biomass differs.

The conversion of biomass into value added chemicals involves the hydrolysis of plant biomass to fermentable sugars. However, to access cellulose a pre-treatment step to remove hemicellulose and lignin is required [20]. Biomass recalcitrance is due to various factors such as the degree of lignification, crystalline structure of cellulose, and the structural heterogeneity and complexity of cell-wall elements and it needs to be overcome for valuable exploitation of lignocellulosic feedstocks [21]. The recalcitrant structure of lignocellulose is disrupted during the pre-treatment process, resulting in lignin sheath breakage, hemicellulose degradation, and a reduction in crystallinity and degree of polymerization of cellulose [22, 23].

There are various types of pre-treatments such as physical methods (such as milling, grinding, irradiation, and sonication); chemical methods (such as alkali, acid, oxidizing agents, organic solvents, ionic liquids, and deep-eutectic solvents); physico-chemical methods (such as steam pretreatment with/without a catalyst, wet-oxidation, and hydrothermolysis); or biological methods (**Table 2**).

The alkali pre-treatment involves the addition of bases to biomass, which increase the internal surface by swelling, and it decreases the polymerization degree and crystallinity, destruction of links between lignin and other polymers, and lignin breakdown [27, 28]. While, acid treatments are known to extract hemicellulose by disrupting the Van der waals force, hydrogen bonds, and covalent bonds that hold the biomass components, which consequently causes the solubilization of hemicellulose and the reduction of cellulose [27]. Steam explosion is another important pretreatment method. Steam explosion is carried out by treating biomass with high-pressure saturated steam, and then the pressure is abruptly lowered, causing the

Biomass	Cellulose (%)	Hemicellulose (%)	Lignin (%)	Reference
Sugarcane bagasse	44	28	24	[10]
Wheat straw	30	50	15	[11]
Rice straw	28–36	23–28	15–23	[12]
Oil palm fronds	45	20	11	[13]
Cotton straw	43	27	18	[14]
Empty fruit bunch	41.1	13.3	25	[15]
Kenaf	32	9.9	40	[15]
Corn stover	34	21	15	[16]
Switch grass	31–38	25–32	17–18	[17]
Sorghum	34	22	19	[16]
Miscanthus	39	20	21	[16]
Elephant grass	36	24	28	[18]
Bamboo	45	24	20	[19]
Corn hobs	45	35	15	[11]

Table 1.
Various feedstock composition analysis.

Pre-treatment methods	Pre-treatment type	Advantages	limitations
Physical methods	Mechanical pulverization	• There is no use of chemicals involved. • Increases pore size and surface area. • Raises the bulk and energy densities. • Reduces the crystallinity of cellulose.	• Energy cost is high. • Require additional pre-treatment steps.
Physicochemical methods	Steam explosion Microwave radiation	• It is responsible for hemicellulose degradation and lignin transformation. • Cost-effective	• A portion of the xylan fraction is disrupted. • Incomplete lignin-carbohydrate matrix disruption. • The production of microorganism-inhibiting compounds.
Chemical methods	Acid hydrolysis	• The hydrolysis of hemicellulose to xylose and other sugars changes the structure of lignin.	• Expensive • Equipment corrosion. • Toxic substance formation.
	Alkaline hydrolysis	• Removes hemicelluloses and lignin. • Increases available surface area.	• Long residence times are necessary. • The formation of irreversible salts and their incorporation into biomass.
Biological methods	Mixed microorganism Adding manure Microaerobic pre-treatment	• Degrades lignin and hemicelluloses. • Low energy requirements.	• Rate of hydrolysis is very low.

Table 2.
Various pre-treatments methods used in biorefineries [24–26].

materials to decompress explosively. Steam explosion is generally initiated at a temperature of 160–260°C (corresponding pressure, 0.69–4.83 MPa) for few seconds to a few minutes before the material is subjected to air pressure [29]. It facilitates the removal of hemicelluloses from the microfibrils exposing the cellulose surface and increasing the enzyme accessibility. Lignin is only partially eliminated during the pretreatment, but it is redistributed on the fiber surfaces as a result of melting and depolymerization/repolymerization processes. Liquid hot water (LHW) treatment is similar to steam explosion, except instead of steam, LHW uses water at extreme temperatures (170–230°C) and pressure (up to 5 MPa). LHW hydrolyzes hemicellulose by releasing its acetyl groups and eliminates lignin, exposing cellulose fibers [30]. **Figure 3** illustrates the necessity of pre-treatments in biomass [31].

These pre-treatments methods can be combined depending on the requirement of the pre-treated material. The pre-treatment method selection is not only based on the yields of the cellulose, hemicellulose, or lignin it was able to extract but there are other factors also to consider, such as the effectiveness of the pre-treatment for a certain post-processing step. For instance, for ethanol production besides sugars, low level of toxic inhibitor compounds such as aldehydes, organic acid,

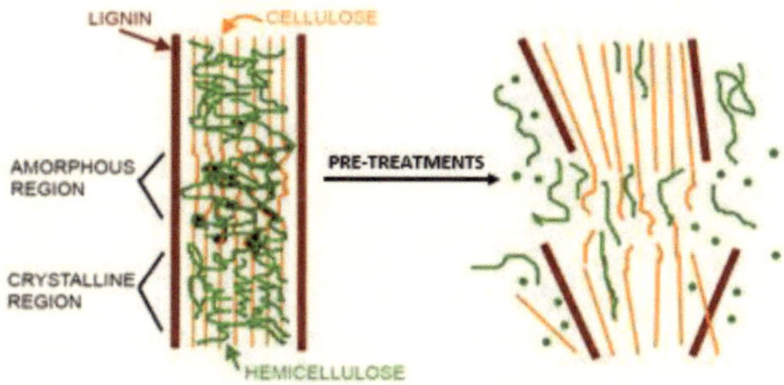

Figure 3.
Illustrates how the pre-treatments detach hemicellulose, cellulose, and lignin from biomass [31].

ketones, phenolic compounds, in the hydrolysate is one of the prerequisite factors [32]. While in case of production of polymers, gels or binders from plant biomass, the mechanical strength and swelling properties of the polymers should be kept in consideration while selecting the appropriate pretreatment method.

4. Platform chemicals

Most chemicals produced in biorefinery are platform chemicals. A platform chemical is a building block which can be converted to a wide range of chemicals or materials. Bomtempo et al. [33] listed ways on how to determine if a compound is a platform chemical:

a. Act as a link in the value chain.

b. Have a flexible chemical structure with functional groups that allow them to produce a variety of derivatives.

c. Have interfaces with varying degrees of openness that allow them to be transformed into a diverse range of derivatives at a reasonable cost. These interfaces are chemical and physical transformation processes that can be integrated or performed by external agents.

d. Possess a flexible chemical structure with functional groups that enable them to synthesize a variety of derivatives.

e. Relate to innovation agents with varying levels of competence and diverse interests which is located at different points along the value chain and are interested in different final products, so that they are structured in innovation ecosystems formed by a set of production chains.

f. Are governed by control and command mechanisms that vary depending on the organizational context. The governance mechanisms should enable the platform leader to capture value in a favorable position.

g. Add value by leveraging scope and scale economies.

Various processes have been used for the production of platform chemicals from cellulose in the biomass based biorefinery. **Table 3** summarize different methods used to produce the platform chemicals from plant biomass.

Building blocks	Processes
1. 5-Hydroxymethyl furfural	Dehydration of fructose
2. Levulinic acid	Acid-catalyzed hydrolysis of glucose
3. Furfural	Dehydration of xylose
4. Succinic acid	Fermentation of glucose
5. Sorbitol	Enzymatic hydrolysis and hydrogenation of glucose
6. Glycerol	• Propylene is synthesized in a variety of ways. • Oil hydrolysis or transesterification of fatty acids/oils • Glycerol can also be produced through yeast fermentation.
7. Ethanol	Thermal gasification with chemical catalytic method, or biochemical fermentation process
8. Lactic acid	Fermentation of glucose
9. Xylitol	Hydrogenation of xylose
10. 2,5-furandicarboxylic acid	HMF oxidation
11. Isoprene	fermentation
12. 3-hydroxypropionic acid	Fermentation of glycerol and oxidation

Table 3.
Biorefinery platform chemicals and production processes [34, 35].

5. Cellulose derived platform chemicals

Among various platform chemicals produced from cellulose, ethanols and acids have already been explored more hence in this chapter the focus is on the less explored platform chemicals which may have more enormous potential.

Levulinic acid (LA): Levulinic acid (LA) also known as 4-oxopentanoic acid with a molecular formula of $C_5H_8O_3$ is one of the twelve valued added chemicals that can be produced from biomass. LA is a ketoacid because it consists of one ketone group (CO) and one carboxylic group (COOH) in its structure [36–38]. Owing to its highly reactive nature LA serves as a versatile platform molecule for fuel additives, resin, herbicides, pharmaceuticals, flavor substances and chemical intermediates with wide potential industrial applications [39–41]. The US Department of Energy selected LA as one of the "12 top value-added compounds from biomass" that may be utilized to produce a variety of high-volume organic chemicals with a wide range of potential industrial uses [39, 40]. **Figure 4** illustrate levulinic acid derivatives applications.

Undergoing condensation reaction, LA produces diphenolic acid (DPA), which is a replacement of bisphenol-A (BPA) used in food containers and consumer products [34]. γ-valerolactone can be produced by hydrogenation of LA, which subsequently can be used as a solvent for lacquers, insecticides, liquid fuel, food addictive and adhesives, and brake fluid. Another LA based product 5-aminolevulinic acid (5-ALA), a porphyrin precursor, is produced via two ways through C4 and C5 pathways. The C4 process, which is present in photosynthetic bacteria, yeast, and human cells, involves the pyridoxal-phosphate-dependent condensation of succinyl-CoA and glycine by ALA synthase (EC 2.3.1.37). The C5 process, found in photosynthetic algae and cyanobacteria, uses glutamate as a co-substratum with ATP and NADPH, and consists of three stages catalyzed by glutamyl-tRNA synthetase, glutamyl-tRNA reductase, and glutamate-1-semialdehyde aminotransferase [42]. 5-ALA is used to

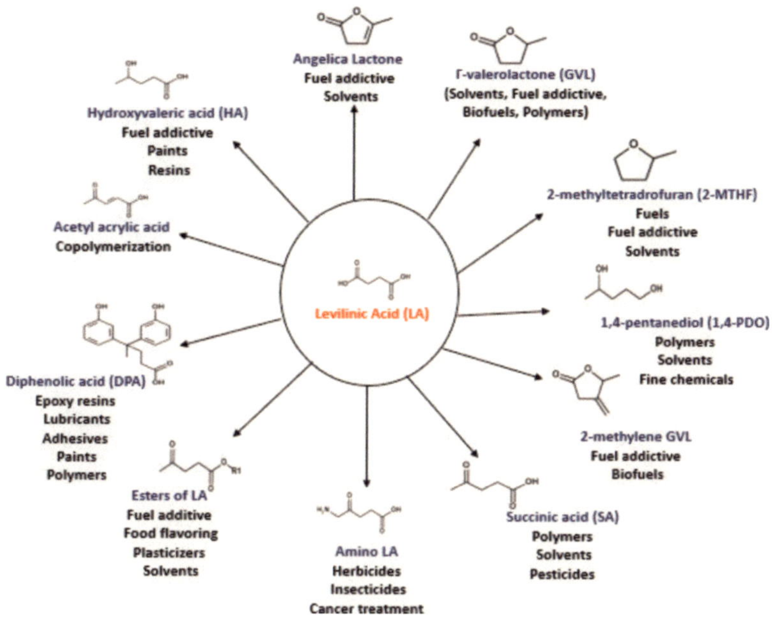

Figure 4.
Illustrate platform levulinic acid applications.

treat actinic keratosis of the face, scalp, and upper extremities, as well as to visualize a glioma, esterification of LA results levulinate esters which are used in the flavoring and fragrance, solvents, plasticizers, or as a blending component in biodiesel [43]. GF Biochemicals located in Caserta, Italy commercialized the production of LA from biomass in 2015 [44].

The conversion of the lignocellulosic biomass to LA requires pre-treatment to remove the lignin and hemicellulose and be left with cellulose which is where LA is produced. To demonstrate effective polymerization of biomass carbohydrate, the catalytic conversion of biomass to LA needs the presence of both Bronsted and Lewis acids [45, 46]. Current commercial production of levulinic acid utilizes sulfuric acid which is corrosive and not environmentally friendly hence more research now focuses on utilizing ionic liquids because they are environmentally friendly. Ionic liquids, which are frequently fluid at room temperature and composed of ionic species, are known as "designer solvents" because their unique characteristics for a given demand may be obtained by suitable alteration of cations or anions. For example, the addition of sulfonic acid (SO_3H) groups and carboxylic acid groups clearly increased their acidities and water solubility, allowing for the development of ecologically acceptable acidic catalysts [47, 48]. Another approach to synthesize LA is one-pot biomass conversion in ionic liquid [49, 50]. However, compared to biomass feed without separation, separated pre-treated biomass will still produce more LA in the following reaction [51].

Succinic acid (SA): Succinic acid (SA) is another important cellulose derived platform chemical. It is member of the C4-dicarboxylic acid family. Succinic acid is also known as butanedioic acid or amber acid, occurs naturally in humans, animals, plants, and microorganisms [52, 53]. Succinic acid is a platform chemical hence it can be used as a precursor to produce varies chemicals namely 1,4-butanediol, tetrahydrofuran, γ-butyrolactone and other bulk chemicals [54], **Figure 5** illustrate succinic acid applications.

Various chemicals can be produced from SA. To mention few, succinate ester is produced via esterification reaction, which is used as a precursor for

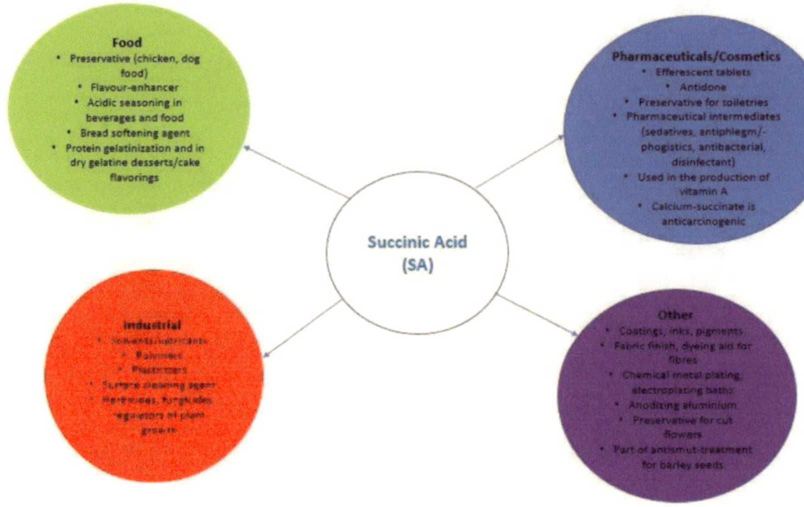

Figure 5.
Illustrate succinic acid applications.

1.4-butanedion, γ-butyrolactone, and tetrahydrofuran [26, 27, 34, 55, 56]. Succinic acid anhydride produced via dehydrogenative cyclization reaction is utilized as a starting material to produce fumaric acid and maleic acid [55]. Coatings, surfactants, dyes, detergents, green solvents, and biodegradable polymers are common SA uses [57]. The worldwide succinic market is predicted to be USD132 million in 2018 and is expected to grow to USD183 million by 2023 [57].

Succinic acid is produced commercially by chemical processes from maleic anhydride, a petroleum-based raw material. Although chemical hydrogenation produces a high yield, purity, and selectivity of succinic acid, it is a complex and costly process with environmental concerns [58]. A more viable method is to use microorganisms to generate succinic acid. Furthermore, bio-based succinic acid has additional environmental benefit using CO_2, a greenhouse gas, as a substrate [59]. Many research on bio-based SA synthesis use pure sugars as substrates. High yields have been produced in these cases using *Anaerobiospirillum succiniciproducens, Asuccinogenes*, modified strains of *Escherichia coli*, and *Mannheimia succiniciproducens*. The biological synthesis of SA is presently being researched as well using lignocellulosic sugars. Succinic acid has been produced by fermentation of corn fiber and sugarcane bagasse [60]. Kuglarz et al. [61] reported on the succinic acid production from rapeseed straw after dilute-acid pre-treatment.

Several companies, such as Myriant, Reverdia, BASF, and BioAmber have commercialized production of succinic acid from glucose [62–64] but SA production from lignocellulosic hydrolysate has yet to be realized at commercial scale [60].

Sorbitol: Sorbitol (D-glucitol, D-sorbitol, D-glucohexane-1,2,3,4,5,6-hexol) is a sugar polyol that is widely utilized in nutrition, cosmetics, and medicinal and industrial purposes. Sorbitol is utilized as a low-calorie sweetener, as a humectant in cosmetics and medicinal goods, and as an intermediate platform for the production of value-added compounds such as 1,4-sorbitan, isosorbide, glycols, l-ascorbic acid, and so on [65]. Sorbitol is one of the most promising platform molecules included in the list of the twelve building block chemicals of highest potential derived from biomass [66, 67]. **Figure 6** illustrate the sorbitol and derivatives applications. Because of sorbitol's rising industrial relevance, there is a lot of interest in upgrading manufacturing methods and looking into novel methods.

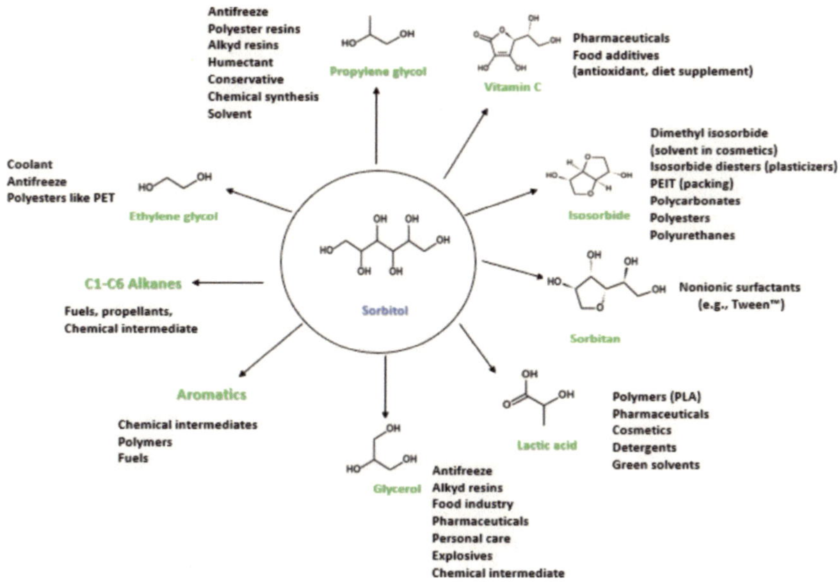

Figure 6.
Illustrate sorbitol and derivatives applications.

The direct conversion of cellulose to sorbitol includes two primary reactions: cellulose hydrolysis to glucose, which is facilitated by the presence of acidic sites, and subsequent hydrogenation of glucose to sorbitol over metal catalysts [68]. One of the major problems in the manufacture of sorbitol is the high cost of processing the raw materials, notably the conversion of cellulose to glucose and the separation process; hence, attempts have been made to accomplish direct conversion of cellulose into sorbitol [69, 70]. Efforts have been undertaken to create a one-pot conversion of cellulose into sorbitol in order to minimize current manufacturing costs. Fukuoka and Dhepe. [71] evaluated the effectiveness of various supporting metal (platinum, Pt, and ruthenium, Ru) catalysts in the conversion of cellulose into sugar alcohols, namely sorbitol and mannitol Pt/g-Al_2O_3 produced the greatest yield (31%) of sugar alcohols with a molar ratio of sorbitol/mannitol of 4:1 or higher.

Zhang et al. [72] produced sorbitol directly from cellulose with the Cu/Al/Fe catalyst in an exceptionally low concentration of phosphoric acid (0.08 percent, w/w), yielding 68.07 percent sorbitol When compared to the fresh catalyst, the catalytic activity of Cu/Al/Fe dropped by 29% in the fourth run. A typical difficulty with direct cellulose-to-sorbitol conversion is the poor sorbitol yield when compared to glucose-to-sorbitol conversion in a commercial process [73]. Zhu et al. [74] also produced sorbitol from cellulose using a sulfonic acid-functionalized silica-supported ruthenium catalyst (Ru/SiO_2–SO_3H), the catalyst was reused five times with a slight decrease in sorbitol yield (up to 61.2%). Sorbitol is commonly industrially produced from the hydrogenation reaction of glucose using metal catalysts, and it global market size is projected to reach USD 2918.1 million by 2026, from USD 2400 million in 2019, at a CAGR of 2.8% during the forecast period 2021–2026 [68, 75–77]. However, the noble metal catalyst is extremely costly for industrial use [74].

Furans*:* Biorefineries can also produce the sugar degradation products such as furans (furfurals and hydroxymethylfurfurals), which is another important platform chemical. **Figure 7** illustrates furfural and 5-hydroxymethylfurfural applications. Furfural is a natural dehydration product of xylose, a monosaccharide often present in high quantities in the hemicellulose portion of lignocellulosic biomass, from which it is almost entirely generated. In theory, any substance having

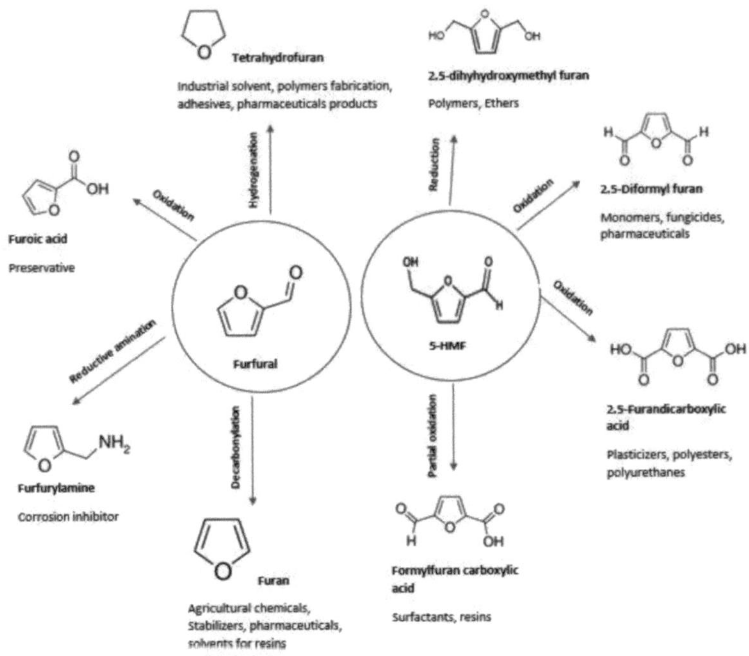

Figure 7.
Applications of furfural and 5-hydroxymethylfurfural derivatives.

a significant quantity of the pentose (five carbon) sugars arabinose and xylose can be used as a raw material for furfural synthesis [78]. It is an aldehyde that consists of heteroaromatic ring. The US Department of Energy [36], which was later revised by Bozell and Petersen [37] identify furfural as one of the top 30 added-value compounds from biomass due to the factors such as manufacturing cost, market price, and function as an intermediary in the manufacture of other important chemicals [36]. These factors are increasing demand, which is anticipated to double between 2014 and 2022 [79]. Furfural is a renewable platform chemical with a diverse chemistry that has the potential to generate new families of bio-based, sustainable chemicals.

These bio-based, sustainable chemicals can be produced via selective hydrogenolysis, reduction, ring opening, aldol condensation reactions [80]. Furfural is a substantial component of bio-oil and is widely used in the production of pharmaceuticals, resins, food additives, fuel additives, and other specialty chemicals [81, 82]. It is a significant component of bio-oil and widely applied in the manufacture of medicines, resins, food additives, fuel additives and other special chemicals [81, 82]. Tetrahydrofurfuryl alcohol, tetrahydrofuran, dihydropyran, acetylfuran, furfurylamine, and furoic acid are other significant furfural-derived compounds. Furfuryl alcohol synthesis followed by acid hydrolysis can also be used to produce levulinic acid from furfural. Furfural is also used in the production of medicines, cosmetics, perfumes, flavors, and resins (the latter produced by condensation with phenol, formaldehyde, acetone, or urea to make thermosetting resins with exceptional physical strength), as well as household cleansers and detergents [79]. While, 5-(hydroxymethyl)furfural (5-HMF) consists of a furan ring with two functional groups namely aldehyde and alcohol group. 5-HMF is utilized for the production of value-added fuels and chemicals (biofuel, solvents, polymers, adhesives, plastic, pesticides, and organic compounds) that were only produced from petroleum-based feedstock [57, 83, 84] 5-HMF is a precursor for the production of levulinic acid,

DOI: http://dx.doi.org/10.5772/intechopen.100022

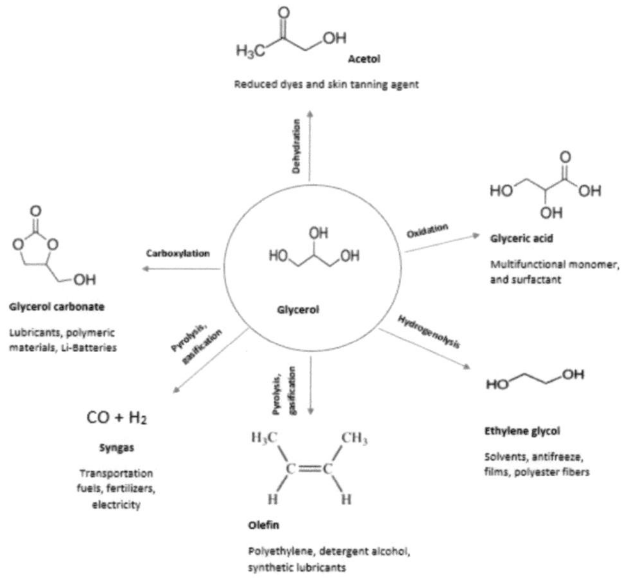

Figure 8.
Illustrates chemicals produced from glycerol with the method used and their applications.

2,5-furan dicarboxylic acid (FDCA), 2,5-diformylfuran, dihydroxymethylfuran, and 5-hydroxy-4-keto-2-pentanoic acid [85].

Glycerol: Glycerol is another important platform chemical that is produced from biomass-based refineries. **Figure 8** illustrates chemicals produced from glycerol and their applications. Glycerol is a versatile carbon source and used as an important raw material for food, pharmaceutical, and cosmetic manufacturing process [86–89].

Glycerol is a common energy-producing food that is widely distributed in food, both naturally and as a GRAS (generally recognized as safe) additive. Glycerol is a solvent that is used in the production of flavors and food colors. It's also used in low-fat food products like cookies as a humectant, plasticizer, emollient, sweetener, and filler.

Glycerol is used in the production of dynamite and propellants (nitroglycerol), cosmetics, candy, liqueurs, printing and copying inks, lubricants, pharmaceuticals (suppositories, cough syrups, elixirs, expectorants, and cardiac medications), personal care products (toothpaste, mouthwashes, skin care products, hair care products, and soaps), and antifreeze. Glycerol can also be utilized to treat glaucoma-related intraocular pressure and cerebral edema [90]. Glycerol is also used to keep textiles malleable, as well as cellophane and high-quality papers flexible and durable. Furthermore, glycerol derivatives include glycerol carbonate, which has several uses in the production of industrial compounds such as glycidol and polymers, coatings, adhesives, and lubricants [34].

Isoprene: Isoprene is a C5 platform chemical that is mostly used for the synthesis of polymers, is another platform chemical to be produced from biomass-based refineries. **Figure 9** illustrate isoprene and derivatives applications. Conventionally, most of the isoprene manufactured is converted into polyisoprene polymer, which is utilized in a wide range of items including footwear, mechanical instruments, medical equipment, rubber tires, and sporting goods [34].

Isoprene is an important platform chemical for synthesizing pesticides, medicines, oil additives, fragrances, and more and is especially important in the rubber production industry. Isoprene is still commercially produced from petroleum-based feedstocks because the utilization of biomass as the feedstock is under investigation.

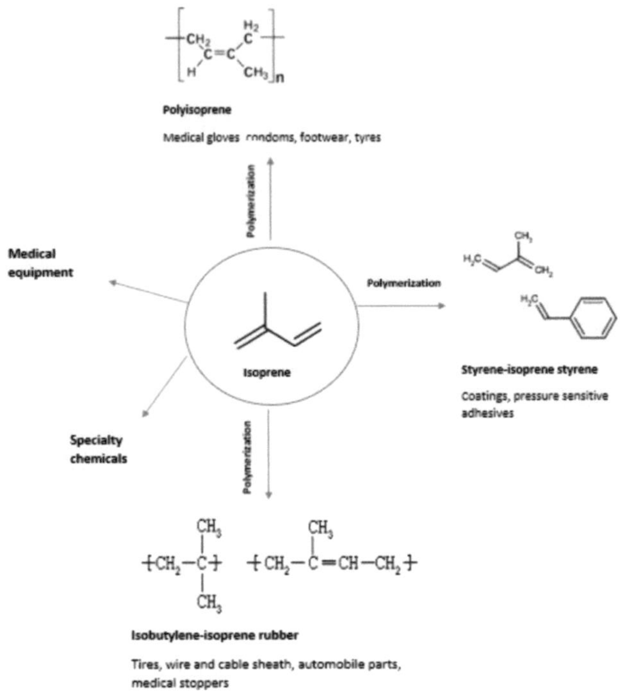

Figure 9.
Illustrates isoprene and derivatives applications.

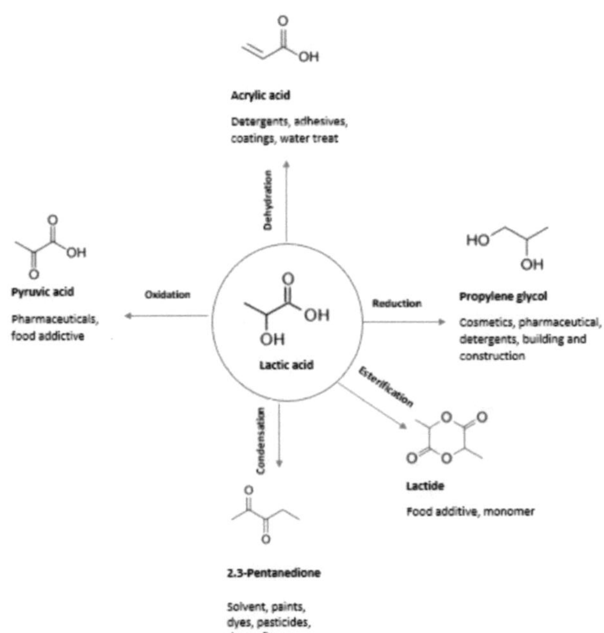

Figure 10.
Illustrates lactic acid and derivatives applications.

Biologically isoprene is produced by fermentation of glucose. Since, the conventional production processes of isoprene is unsustainable, and the chemical synthesis processes could cause serious environmental problems [91].

Lactic Acid: Lactic acid is another platform chemical which is produced from biorefinery. It is commercially produced via the fermentation of various sugars namely glucose, sucrose, or lactose [92]. **Figure 10** shows lactic acid and derivatives applications. Direvo industrial biotechnology produces lactic acid from biomass in a single step using consolidated bioprocessing technology [93]. Lactic acid as a platform chemical, produces chemicals such as lactate esters by esterification which is used as a green solvent. The reduction of lactic acid produces propylene glycol, which is used in food industry as a humectant, solvent, and preservative. Besides, lactic acid can also be used in medicine as a drug stabilizer and a solvent, in cosmetics as a humectant and can also be used in E-cigarettes [94].

6. Conclusions

Biomass based biorefineries are the best sustainable alternative to the fossil fuel crisis and environment protection. There are varies types of biomass with varying cellulose, hemicellulose, and lignin content. To extract cellulose from biomass a pre-treatment is necessary due to biomass components being recalcitrant. The common pre-treatments have disadvantages and limitations, and it is necessary to understand that the pre-treatment cannot be selected based on the yield of cellulose only because there are other effect caused by the pre-treatment which may be in conflict with the next reaction step to value-added products. This chapter have made an attempt to throw some light on the significance of the use of cellulose as a potential feedstock for various platform chemicals. Although the production of such platform chemicals using green chemistry approach is in very nascent phase however these processes have shown immense potential. Both government and industries should switch their focus towards the development of environmentally friendly and efficient process for such platform chemicals.

Acknowledgements

The Authors wish to show gratitude to Durban University of Technology and L'Oréal-UNESCO for Women in Science Sub-Saharan Africa Regional fellowship for funding.

Conflict of interest

The authors declare no conflict of interest.

Author details

Lethiwe D. Mthembu[1]*, Rishi Gupta[2] and Nirmala Deenadayalu[1]

1 Department of Chemistry, Durban University of Technology, Durban, South Africa

2 Anton Paar India Pvt Ltd., Udyog Vihar, Gurugram, Haryana, India

*Address all correspondence to: lethiwem@dut.ac.za

References

[1] Pulicharla R, Lonappan L, Brar SK, Verma M. Production of renewable C5 platform chemicals and potential applications. In Platform Chemical Biorefinery. Elsevier. 2016; 201-216. DOI: 10.1016/B978-0-12-802980-0.00011-0

[2] Ng DKS, Ng KS, Ng R T L. Integrated Biorefineries. Encyclopedia of Sustainable Technologies. 2017; 299-314. DOI:10.1016/b978-0-12-409548-9.10138-1

[3] https://www.bccresearch.com/market-research/energy-and-resources/biorefinery-products-markets-report.html, accessed June 2021

[4] Galbe M, Wallberg O. Pretreatment for biorefineries: a review of common methods for efficient utilisation of lignocellulosic materials. Biotechnology for Biofuels. 2019; 12: 1-26. DOI:10.1186/s13068-019-1634-1

[5] Bai FW, Yang S, Ho NW. Fuel Ethanol Production from Lignocellulosic Biomass, Ed Murray Moo-Young, Comprehensive Biotechnology (Third Edition), Pages 49-65, (2019). DOI: 10.1016/B978-0-444-64046-8.00150-6

[6] Mudgil, D. (2017). The Interaction Between Insoluble and Soluble Fiber. Dietary Fiber for the Prevention of Cardiovascular Disease, 35-59. doi:10.1016/b978-0-12-805130-6.00003-3

[7] De Carvalho, K. C. C., Montoro, S. R., Cioffi, M. O. H., & Voorwald, H. J. C. (2016). Polyhydroxyalkanoates and Their Nanobiocomposites With Cellulose Nanocrystals. Design and Applications of Nanostructured Polymer Blends and Nanocomposite Systems, 261-285. doi:10.1016/b978-0-323-39408-6.00012-1

[8] Gupta, P.K., Raghunath, S.S., Prasanna, D.V., Venkat, P., Shree, V., Chithananthan, C., Choudhary, S., Surender, K. and Geetha, K., 2019. An update on overview of cellulose, its structure and applications. *Cellulose*, pp.846-1297.

[9] https://biocyclopedia.com/index/biotechnology/biotechnology_and_environment/biomass_a_renewable_source_of_energy/biotech_biomass_composition_of_biomass.php, accessed March 2021

[10] Schmidt LM, Mthembu LD, Reddy P, Deenadayalu N, Kaltschmitt M, Smirnova I. Levulinic acid production integrated into a sugarcane bagasse based biorefinery using thermal-enzymatic pretreatment. Industrial Crops Products. 2017; 99: 172-178. DOI: 1016/j.indcrop.2017.02.010

[11] Jorgensen H, Kristensen JB, Felby C. Enzymatic conversion of lignocellulose into fermentable sugars: Challenges and opportunities. Biofuels, Bioproducts and Biorefining. 2007; 1: 119-134. DOI: 10.1002/bbb.4

[12] Kaur A, Kuhad RC. Valorization of rice straw for ethanol production and lignin recovery using combined acid-alkali pre-treatment. BioEnergy Research. 2019; 12: 570-582. DOI: 10.1007/s12155-019-09988-3

[13] Ramli NAS, Amin NAS. Optimization of biomass conversion to levulinic acid in acidic ionic liquid and upgrading of levulinic acid to ethyl levulinate. BioEnergy Research. 2017; 10: 50-63. DOI: 10.1007/s12155-016-9778-3

[14] Yang Z, Kang H, Guo Y, Zhuang G, Bai Z, Zhang H, Feng C, Dong Y. Dilute-acid conversion of cotton straw to sugars and levulinic acid via 2-stage

hydrolysis. Industrial Crops and Products. 2013; 46: 205-209. DOI: 10.1016/j.indcrop.2013.01.031

[15] Ya'aini N, Amin NAS, Asmadi M. Optimization of levulinic acid from lignocellulosic biomass using a new hybrid catalyst. Bioresource Technology. 2012; 116:58-65. DOI: 10.1016/j.biortech.2012.03.097

[16] Williams CL, Emerson RM, Tumuluru JS. Biomass compositional analysis for conversion to renewable fuels and chemicals. Biomass volume estimation and valorization for energy. 2017; 251-270. DOI: 10.5772/65777

[17] David K, Ragauskas AJ. Switchgrass as an energy crop for biofuel production: A review of its ligno-cellulosic chemical properties. Energy & Environmental Science. 2010; 3:1182. DOI:10.1039/b926617h

[18] Scholl AL, Menegol D, Pitarelo AP, Fontana RC, Zandoná Filho A, Ramos LP, Dillon AJP, Camassola M. Elephant grass pretreated by steam explosion for inducing secretion of cellulases and xylanases by Penicillium echinulatum S1M29 solid-state cultivation. Industrial Crops and Products. 2015; 77:97-107. DOI: 10.1016/j.indcrop.2015.08.051

[19] Li X, Sun C, Zhou B, He Y. Determination of hemicellulose, cellulose and lignin in moso bamboo by near infrared spectroscopy. Scientific reports. 2015; 5:1-11. DOI: 10.1038/srep17210

[20] Singh S, Cheng G, Sathitsuksanoh N, Wu D, Varanasi P, George A, Balan V, Gao X, Kumar R, Dale BE, Wyman CE, Simmons BA. Comparison of different biomass pretreatment techniques and their impact on chemistry and structure. Front Energy Res Bioenergy Biofuels. 2015; 2:1-12. DOI:10.3389/fenrg.2014.00062

[21] Guerriero G, Hausman JF, Strauss J, Ertan H, Siddiqui KS. Lignocellulosic biomass: biosynthesis, degradation, and industrial utilization. Engineering in Life Sciences. 2016; 16:1-16. DOI: 10.1002/elsc.201400196

[22] Loow YL, Wu TY, Tan KA, Lim YS, Siow LF, Md Jahim J, Mohammad AW, Teoh WH. Recent advances in the application of inorganic salt pretreatment for transforming lignocellulosic biomass into reducing sugars. Journal of Agricultural and Food Chemistry. 2015; 63: 8349-8363. DOI: 10.1021/acs.jafc.5b01813

[23] Chen H, Liu J, Chang X, Chen D, Xue Y, Liu P, Lin H, Han S. A review on the pretreatment of lignocellulose for high-value chemicals. Fuel Processing Technology. 2017;160: 196-206. DOI: 10.1016/j.fuproc.2016.12.007

[24] Kumar P, Barrett DM, Delwiche MJ, Stroeve P. Methods for pretreatment of lignocellulosic biomass for efficient hydrolysis and biofuel production. Industrial & engineering chemistry research. 2009; 48:3713-3729. DOI: 10.1021/ie801542g

[25] Amin FR, Khalid H, Zhang H, u Rahman S, Zhang R, Liu G, Chen C. Pretreatment methods of lignocellulosic biomass for anaerobic digestion. Amb Express. 2017; 7: 1-12. DOI: 10.1186/s13568-017-0375-4

[26] Anukam A, Berghel J. Biomass Pretreatment and Characterization: A Review. Biomass. 2020. DOI: 10.5772/intechopen.93607

[27] Amin FR, Khalid H, Zhang H, u Rahman S, Zhang R, Liu G, Chen C. Pretreatment methods of lignocellulosic biomass for anaerobic digestion. Amb Express. 2017; 7: 1-12. DOI: 10.1186/s13568-017-0375-4

[28] Kaur A, Kuhad, RC. Valorization of rice straw for ethanol production and

lignin recovery using combined acid-alkali pre-treatment. BioEnergy Research. 2019; 12: 570-582. DOI: 10.1007/s12155-019-09988-3

[29] Kumar P, Barrett DM, Delwiche MJ, Stroeve P. Methods for pretreatment of lignocellulosic biomass for efficient hydrolysis and biofuel production. Industrial & engineering chemistry research. 2009; 48:3713-3729. DOI: 10.1021/ie801542g

[30] Baruah J, Nath BK, Sharma R, Kumar S, Deka RC, Baruah DC, Kalita E. Recent trends in the pretreatment of lignocellulosic biomass for value-added products. Frontiers in Energy Research. 2018; 6:141. DOI: 10.3389/fenrg.2018.00141

[31] Pranjali M, Dorin, B. Advances in Biomass Pretreatment and Cellulosic Bioethanol Production Using Microwave Heating. 2017; 173-180. DOI: 10.18690/978-961-286-048-6.18.

[32] Galbe M, Wallberg O. Pretreatment for biorefineries: A review of common methods for efficient utilisation of lignocellulosic materials. Biotechnology for biofuels. 2019; 12: 1-26. DOI: 10.1186/s13068-019-1634-1

[33] Bomtempo JV, Chaves Alves F, de Almeida Oroski F. Developing new platform chemicals: what is required for a new bio-based molecule to become a platform chemical in the bioeconomy? Faraday Discussions. 2017; 202: 213-225. DOI:10.1039/c7fd00052a

[34] Takkellapati S, Li T, Gonzalez MA. An overview of biorefinery-derived platform chemicals from a cellulose and hemicellulose biorefinery. Clean technologies and environmental policy. 2018; 20: 1615-1630. DOI: 10.1007/s10098-018-1568-5

[35] Bagnato G, Iulianelli A, Sanna A, Basile A. Glycerol production and transformation: a critical review with particular emphasis on glycerol reforming reaction for producing hydrogen in conventional and membrane reactors. Membranes. 2017; 7:17. DOI: 10.3390/membranes7020017

[36] Werpy T, Petersen G. Top value-added chemicals from biomass: volume I--results of screening for potential candidates from sugars and synthesis gas (No. DOE/GO-102004-1992). National Renewable Energy Lab., Golden, CO (US). 2004

[37] Bozell JJ, Petersen GR. Technology development for the production of biobased products from biorefinery carbohydrates—the US Department of Energy's "Top 10" revisited. Green Chemistry. 2010; 12: 539-554. DOI: 10.1039/B922014C

[38] Serrano-Ruiz JC, Wang D, Dumesic JA. Catalytic upgrading of levulinic acid to 5-nonanone. Green Chemistry. 2010; 12: 574-577. DOI: 10.1039/B923907C

[39] Weingarten R, Conner WC, Huber GW. Production of levulinic acid from cellulose by hydrothermal decomposition combined with aqueous phase dehydration with a solid acid catalyst. Energy & Environmental Science. 2012; 5: 7559. DOI:10.1039/c2ee21593d

[40] Bozell JJ. Connecting Biomass and Petroleum Processing with a Chemical Bridge. Science. 2010; 329: 522-523. DOI:10.1126/science.1191662

[41] Shen Y, Sun JK, Yi YX, Wang B, Xu F, Sun RC. One-pot synthesis of levulinic acid from cellulose in ionic liquids. Bioresource Technology. 2015; 192: 812-816. DOI: 10.1016/j.biortech.2015.05.080

[42] Shin JA, Kwon YD, Kwon OH, Lee HS, Kim P. 5-Aminolevulinic acid biosynthesis in Escherichia coli coexpressing NADP-dependent malic

enzyme and 5-aminolevulinate synthase. Journal of microbiology and biotechnology. 2007; 17: 1579-1584. PMID: 18062242.

[43] Mthembu LD, Lokhat D, Deenadayalu N. Esterification of levulinic acid to ethyl levulinate: optimization of process conditions using commercial levulinic acid and extension to the use of levulinic acid derived from depithed sugarcane bagasse. Biomass Conversion and Biorefinery. 2021; 1-10. DOI: 10.1007/s13399-021-01632-5

[44] http://www.gfbiochemicals.com, accessed February 2021

[45] Tiong YW, Yap CL, Gan S, Yap WSP. One-pot conversion of oil palm empty fruit bunch and mesocarp fiber biomass to levulinic acid and upgrading to ethyl levulinate via indium trichloride-ionic liquids. Journal of Cleaner Production. 2017; 168: 1251-1261. DOI: 10.1016/j.jclepro.2017.09.050

[46] Ramli NAS, Amin NAS. Optimization of oil palm fronds conversion to levulinic acid using Fe/HY zeolite catalyst. Sains Malaysiana. 2015; 44: 883-891. DOI: 10.17576/jsm-2015-4406-15

[47] Shen Y, Sun JK, Yi YX, Wang B, Xu F, Sun RC. One-pot synthesis of levulinic acid from cellulose in ionic liquids. Bioresource Technology. 2015; 192: 812-816. DOI: 10.1016/j.biortech.2015.05.080

[48] Cole AC, Jensen JL, Ntai I, Tran KLT, Weaver KJ, Forbes DC, Davis JH. Novel bronsted acidic ionic liquids and their use as dual solvent-catalysts. Journal of the American Chemical Society. 2002; 124: 5962-5963. DOI: 10.1021/ja026290w

[49] Tiong YW, Yap CL, Gan S, Yap WSP. Optimisation studies on the conversion of oil palm biomass to levulinic acid and ethyl levulinate via indium trichloride-ionic liquids: a response surface methodology approach. Industrial Crops Products. 2019; 128: 221-234. DOI: 10.1016/j.indcrop.2018.10.087

[50] Ramli NAS, Amin NAS. Optimization of biomass conversion to levulinic acid in acidic ionic liquid and upgrading of levulinic acid to ethyl levulinate. BioEnergy Research. 2016; 1-14. DOI: 10.1007/s12155-016-9778-3

[51] Ramli NAS, Amin NAS. Catalytic hydrolysis of cellulose and oil palm biomass in ionic liquid to reducing sugar for levulinic acid production. Fuel Processing Technology. 2014; 128: 490-498. DOI: 10.1016/j.fuproc.2014.08.011

[52] Zeikus JG, Jain MK, Elankovan P. Biotechnology of succinic acid production and markets for derived industrial products. Applied and Environmental Microbiology. 1999; 51: 545-552. DOI: 10.1007/s002530051431

[53] Jiang M, Xu R, Xi YL, Zhang JH, Dai WY, Wan YJ, Chen KQ, Wei P. Succinic acid production from cellobiose by Actinobacillus succinogenes. Bioresource Technology. 2013; 135:469-474.

[54] McKinlay JB, Vieille C, Zeikus JG. Prospects for a bio-based succinate industry. Applied microbiology and biotechnology. 2007; 76: 727-740. DOI: 10.1007/s00253-007-1057-y

[55] Delhomme C, Weuster-Botz D, Kuhn FE. Succinic acid from renewable resources as a C4 building-block chemical—a review of the catalytic possibilities in aqueous media. Green Chemistry. 2009; 11:13-26. DOI: 10.1039/B810684C

[56] Luque R, Clark JH, Yoshida K, Gai PL. Efficient aqueous hydrogenation of biomass platform molecules using supported metal nanoparticles on

Starbons. Chemical Communications. 2009; 35:5305-5307. DOI: 10.1039/B911877B

[57] Trivedi J, Bhonsle AK, Atray N. Processing food waste for the production of platform chemicals. In Refining Biomass Residues for Sustainable Energy and Bioproducts (pp. 427-448). Academic Press. 2020. DOI: 10.1016/B978-0-12-818996-2.00019-3

[58] Vasudevan D. Reduction of maleic acid at a Ti / ceramic TiO_2 cathode. Journal of Applied Electrochemistry. 1995; 25: 176-178. DOI: 10.1007/BF00248176

[59] Pasma SA, Daik R, Maskat MY. Production of succinic acid from oil palm empty fruit bunch cellulose using *Actinobacillus succinogenes*. 2013; 1571:753-759. DOI:10.1063/1.4858745

[60] Salvachua D, Smith H, John PCS, Mohagheghi A, Peterson DJ, Black BA, Dowe N, Beckham GT. Succinic acid production from lignocellulosic hydrolysate by Basfia succiniciproducens. Bioresource Technology. 2016; 214: 558-566. DOI: 10.1016/j.biortech.2016.05.018

[61] Kuglarz M, Alvarado-Morales M, Dąbkowska K, Angelidaki I. Integrated production of cellulosic bioethanol and succinic acid from rapeseed straw after dilute-acid pretreatment. Bioresource Technology. 2018; 265:191-199. DOI: 10.1016/j.biortech.2018.05.099

[62] Erickson B, Nelson JE, Winters P. Perspective on opportunities in industrial biotechnology in renewable chemicals. Biotechnology Journal. 2012; 7: 176-185. DOI: 10.1002/biot.201100069

[63] Ge X, Chang C, Zhang L, Cui S, Luo X, Hu S, Qin Y, Li Y. Conversion of Lignocellulosic Biomass Into Platform Chemicals for Biobased Polyurethane Application. Advances in Bioenergy. 2018; 3: 161-213. DOI: 10.1016/bs.aibe.2018.03.002

[64] Cok B, Tsiropoulos I, Roes AL, Patel MK. Succinic acid production derived from carbohydrates: An energy and greenhouse gas assessment of a platform chemical toward a bio-based economy. Biofuels, Bioproducts and Biorefining. 2014; 8: 16-29. DOI: 10.1002/bbb.1427

[65] García B, Moreno J, Morales G, Melero JA, Iglesias J. Production of Sorbitol via Catalytic Transfer Hydrogenation of Glucose. Applied Sciences. 2020; 10:1843. DOI: 10.3390/app10051843

[66] Ochoa-Gómez JR, Roncal T. Production of Sorbitol from Biomass. In: Fang Z, Smith Jr. R, Qi X. (eds) Production of Platform Chemicals from Sustainable Resources. Biofuels and Biorefineries. Springer, Singapore. 2017; 265-309. DOI:10.1007/978-981-10-4172-3_9

[67] Yabushita M, Kobayashi H, Fukuoka A. Catalytic transformation of cellulose into platform chemicals. Applied Catalysis B: Environmental. 2014; 145: 1-9. DOI: 10.1016/j.apcatb.2013.01.052

[68] Zada B, Chen M, Chen C, Yan L, Xu Q, Li W, Guo Q, Fu Y. Recent advances in catalytic production of sugar alcohols and their applications. Science China Chemistry. 2017; 60: 853-869. DOI: 10.1007/s11426-017-9067-1

[69] Shrotri A, Kobayashi H, Fukuoka A. Cellulose Depolymerization over Heterogeneous Catalysts. Accounts of Chemical Research. 2018; 51: 761-768. DOI: 10.1021/acs.accounts.7b00614

[70] Zhang J, Li JB, Wu SB, Liu Y. Advances in the catalytic production and utilization of sorbitol. Industrial & Engineering Chemistry Research. 2013; 52: 11799-11815. DOI: 10.1021/ie4011854

[71] Fukuoka A, Dhepe PL. Catalytic conversion of cellulose into sugar alcohols. Angewandte Chemie International Edition in English. 2006; 45: 5161-5163. DOI: 10.1002/anie.200601921

[72] Zhang J, Wu SB, Liu Y. Direct conversion of cellulose into sorbitol over a magnetic catalyst in an extremely low concentration acid system. Energy Fuel. 2014; 28: 4242-4246. DOI: 10.1021/ef500031w

[73] Ge X, Chang C, Zhang L, Cui S, Luo X, Hu S, Qin Y, Li Y. Conversion of Lignocellulosic Biomass Into Platform Chemicals for Biobased Polyurethane Application. Advances in Bioenergy. 2018; 3: 161-213. DOI: 10.1016/bs.aibe.2018.03.002

[74] Zhu W, Yang H, Chen J, Chen C, Guo L, Gan H, Zhao X, Hou Z. Efficient hydrogenolysis of cellulose into sorbitol catalyzed by a bifunctional catalyst. Green Chemistry. 2014; 16: 1534-1542. DOI: 10.1039/c3gc41917g

[75] Rey-Raap N, Ribeiro LS, Órfão JJM, Figueiredo JL, Pereira MFR. Catalytic conversion of cellulose to sorbitol over Ru supported on biomass-derived carbon-based materials. Applied Catalysis B: Environmental. 2019; 256: 117826. DOI: 10.1016/j.apcatb.2019.117826

[76] Besson M, Gallezot P, Pinel C. Conversion of biomass into chemicals over metal catalysts. Chemical reviews. 2014; 114: 1827-1870. DOI: 10.1021/cr4002269

[77] https://www.prnewswire.com/news-releases/sorbitol-market-size-to-reach-usd-2918-1-million-by-2026-at-cagr-2-8--valuates-reports-301246413.html, accessed July 2021

[78] Cai CM, Zhang T, Kumar R, Wyman CE. Integrated furfural production as a renewable fuel and chemical platform from lignocellulosic biomass. Journal of Chemical Technology & Biotechnology. 2014; 89: 2-10. DOI: 10.1002/jctb.4168

[79] Peleteiro S, Rivas S, Alonso JL, Santos V, Parajó JC. Furfural production using ionic liquids: A review. Bioresource Technology. 2016; 2: 181-191. DOI: 10.1016/j.biortech.2015.12.017

[80] Shinde SH, Hengne A. Rode CV. Lignocellulose-derived platform molecules. Biomass, Biofuels, Biochemicals. 2020; 1-31. DOI:10.1016/b978-0-444-64307-0.00001-9

[81] Lu Q, Wang Z, Dong CQ, Zhang ZF, Zhang Y, Yang YP, Zhu XF. Selective fast pyrolysis of biomass impregnated with $ZnCl_2$: Furfural production together with acetic acid and activated carbon as by-products. Journal of Analytical and Applied Pyrolysis. 2011; 91: 273-279. DOI: 10.1016/j.jaap.2011.03.002

[82] Zhang H, Liu X, Lu M, Hu X, Lu L, Tian X, Ji J. Role of Brønsted acid in selective production of furfural in biomass pyrolysis. Bioresource Technology. 2014; 169: 800-803. DOI: 10.1016/j.biortech.2014.07.053

[83] Mukherjee A, Dumont MJ, Raghavan V. Review: sustainable production of hydroxymethylfurfural and levulinic acid: challenges and opportunities. Biomass Bioenergy. 2015; 72:143-183. DOI: 10.1016/j.biombioe.2014.11.007

[84] Bhaumik P, Dhepe PL. Solid acid catalyzed synthesis of furans from carbohydrates. Catalysis Reviews. 2016; 58: 36-112. DOI: 10.1080/01614940.2015.1099894

[85] Rosatella AA, Simeonov SP, Frade RFM, Afonso CAM, 2011. 5-Hydroxymethylfurfural (HMF) as a building block platform: biological properties, synthesis and synthetic

applications. Green Chemistry. 2011; 13:754. DOI: 10.1039/c0gc00401d

[86] Luo X, Ge X, Cui S, Li Y. Value-added processing of crude glycerol into chemicals and polymers. Bioresource Technology. 2016; 215: 144-154. DOI: 10.1016/j.biortech.2016.03.042

[87] Guerrero-Perez MO, Juana MR, Jorge B, Jose RM, Tomas C. Recent inventions in glycerol transformations and processing. Recent Patents on Engineering. 2009; 2: 11-21. DOI: 10.2174/1874478810902010011

[88] Chozhavendhan S, Praveenkumar R, Sivarathanakumar S, Raja Sathendra E, Bharathiraja B, Jayakumar M. Comparative study on Candida sp for the production glycerol. International Journal of ChemTech Research. 2014; 6: 5058-5063.

[89] Chozhavendhan S, Karthiga Devi G, Jayamuthunagai J, Bharathiraja B, Praveen kumar R, Raman JK. Conversion of Glycerol to Valuable Products. In: Pogaku R. (eds) Horizons in Bioprocess Engineering. Springer, Cham. 2019. DOI: 10.1007/978-3-030-29069-6_8

[90] Wurzel KA. Glycerol. Encyclopedia of Toxicology. 2005; 449-451. DOI:10.1016/b0-12-369400-0/00457-9

[91] Li M, Nian R, Xian M, Zhang H. Metabolic engineering for the production of isoprene and isopentenol by Escherichia coli. Applied microbiology and biotechnology. 2018; 102:7725-7738. DOI: 10.1007/s00253-018-9200-5

[92] Ghaffar T, Irshad M, Anwar Z, Aqil T, Zulifqar Z, Tariq A, Kamran M, Ehsan N, Mehmood S. Recent trends in lactic acid biotechnology: a brief review on production to purification. Journal of radiation research and applied Sciences. 2014; 7:222-229. DOI: 10.1016/j.jrras.2014.03.002

[93] Direvo press release. http://www.direvo.com/uploads/media/DIREVO-PressRelease_No6_2013.pdf Accessed June 2017

[94] https://www.chemicals.co.uk/blog/5-uses-of-propylene-glycol, accessed May 2021

Chapter 9

Cellulose as a Natural Emulsifier: From Nanocelluloses to Macromolecules

Carolina Costa, Bruno Medronho, Björn Lindman, Håkan Edlund and Magnus Norgren

Abstract

During the last decade, cellulose structural features have been revisited, with particular focus on its structural anisotropy (amphiphilicity) and interactions determining its recalcitrance to dissolution. Evidences for cellulose amphiphilicity are patent, for instance, in its capacity to adsorb at oil–water interfaces, thus being capable of stabilizing emulsions. This behavior is observable in all its forms, from cellulose nanoparticles to macromolecules. This chapter is divided into two main parts; first, the fundamentals of emulsion formation and stabilization will be introduced, particularly focusing on the role of natural emulsifiers. Secondly, the emerging role of cellulose as a natural emulsifier, where the ability of cellulose to form and stabilize emulsions is revisited, from cellulose nanoparticles (Pickering-like effect) to macromolecules (i.e., cellulose derivatives and native molecular cellulose).

Keywords: Nanocelluloses, regenerated cellulose, cellulose derivatives, native molecular cellulose, amphiphilicity, oil–water interface, emulsifiers, adsorption, emulsions, stabilization

1. Introduction

Every year, hundred billion tons of cellulose are produced by nature from various biomass sources, making this biopolymer an ultimate platform for developing sustainable applications on an industrial scale [1]. The increased environmental awareness due to global climate changes has pushed cellulose science to advance rapidly, and thus cellulose is expected to continue playing a central role in the emergent bio-economies and biorefineries. Cellulose extraction and purification rely on fairly simple, scalable, and efficient isolation techniques, and cellulose can be further modified and shaped into different colloidal and macroscopic forms, showing very different features [2–5].

Emulsions are among the most important colloids in everyday life, and have multiple uses, from technical applications, such as paints and coatings, to life science applications, such as foods, pharmaceuticals, and cosmetics. Emulsions can encapsulate and protect sensitive ingredients, adjust appearance, taste and sensorial properties, and facilitate application, spreading and drying. They can also serve as an intermediator for efficient oil extraction, polymerization reactions, and the production of microcapsules and lightweight foams [6]. Their structural and functional properties are vast and therefore, the emulsifiers and stabilizers must be carefully selected according to the needs. With the growing global demand for sustainable and "clean-label" products, industries are actively seeking to replace synthetic emulsifiers by new alternatives. Finding natural ingredients, with minimal modification requirements (without compromising environment and human health) and still being capable of achieving a competitive performance to the optimized synthetic options is highly desirable but challenging. In this respect, cellulose has the potential to become a key star player in emulsion systems. In addition to its natural, non-toxic, biodegradable, and renewable nature, it is also a versatile source of natural emulsifiers. From cellulose derivatives, used since the early 20th century, to the more recently explored native forms of cellulose, including nanocelluloses, and the native cellulose itself, either molecular or in the form of polymeric particles and microgels, all its different morphological forms have shown an intrinsic amphiphilic character by self-assembling at oil–water interfaces.

2. Emulsion formation and stabilization by natural polymers and particles

Emulsions are multiphasic systems of at least three main components, the oil phase, the water phase, and the emulsifier. One of the phases is dispersed into the other in the form of droplets that are stabilized by a key compound, an emulsifier. Depending on the dispersed phase, we can have either oil-in-water (o/w) or water-in-oil (w/o) emulsions, and the type of emulsion formed mainly depends on the solubility properties of the emulsifier. According to Bancroft's rule, o/w emulsions are formed when the emulsifier has a preference for water whereas the opposite applies for w/o emulsions [7]. When an emulsion is formed, a large interfacial area is created between the two phases, generating an increased energy in relation to the interfacial tension between oil and water. Therefore, emulsions seek to minimize the energy used to create such large interfacial area and break down over time by the combination of different instability mechanisms, such as, creaming, flocculation, coalescence and Ostwald ripening [7]. The role of the emulsifier is to reduce the interfacial tension and form a "protective layer" through its adsorption on the droplets surface, thus facilitating not only the formation of the droplets but also preventing/minimizing their re-association. Amphiphilic molecules and insoluble particles have both been employed as emulsifiers (**Figure 1**). Small surfactant molecules are usually good emulsifiers. Nevertheless, they are often not particularly well suited to provide long-term stability; this is because they are in dynamic equilibrium with the bulk medium. In this case, often, a stabilizer is required to achieve sufficient kinetic stability for the required shelf-life of a certain product. Polymers

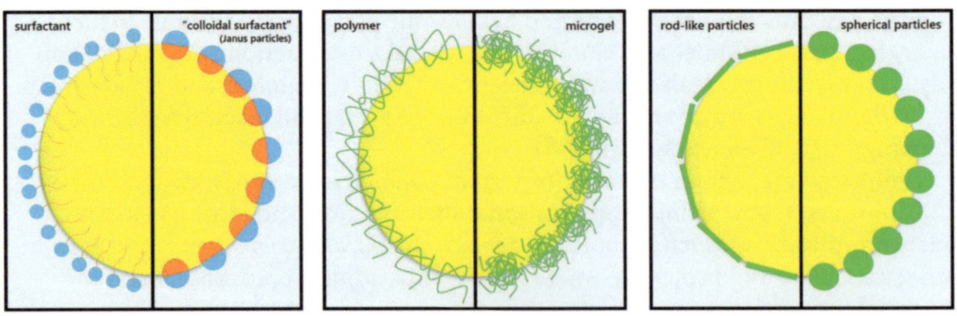

Figure 1.
Emulsifiers: Surfactants, polymers and particles. Differences in scaling are not considered.

are often applied as stabilizers in oil-in-water (o/w) emulsions, and they can act either via the reinforcement of the stabilizing layer, co-acting with the emulsifier at the interface, or via the viscosity enhancement of the continuous phase, thus reducing droplet mobility [8]. Certain amphiphilic polymers and particles may act as both emulsifiers and stabilizers. Good examples are surface-active polysaccharides, such as gum Arabic, pectin, galactomannans and modified starches and celluloses [8–11]. These polymers provide strong steric repulsions driven by the entropic penalty when polymer segments from two droplets start to entangle, since conformational rearrangements are hindered due to their high molecular weight [12].

Another type of stabilization is provided by insoluble particles, often called Pickering stabilization; rigid particles, Janus particles and microgels, have been described as Pickering emulsifiers [13–15]. The amphiphilicity of a typical Pickering emulsifier (rigid particles) is usually described in terms of surface wettability, which is measured by the three-phase contact angle of a particle adsorbed at an oil–water interface. Both o/w and w/o emulsions can be formed depending on the particle wettability and whether the particles are predominantly hydrophilic or hydrophobic [13]. In agreement with Bancroft's rule, the interface tends to bend towards the more poorly wetted liquid, and this becomes the dispersed phase. Pickering particles adsorb irreversibly at the oil–water interfaces due to the high-binding energy per particle, forming an effective mechanical barrier against coalescence; they may also inhibit lipid oxidation due to the thick interfacial layers formed [10, 11]. This is an important feature in what concerns food and pharmaceutical applications where polyunsaturated lipids are involved. Their double bonds are prone to oxidation leading to the deterioration of the products by the formation of rancid flavors and, eventually, toxic by-products [16]. The most widely used bioparticles are derived from biopolymers, such as cellulose, chitin and chitosan, starch and modified starches, lignin and proteins [17–21]. Bioparticles may vary widely in shape, size, aspect ratio and morphology, implying that their mechanistic behavior considerably deviates from that of both the solid sphere and the flexible polymer [22]. Nevertheless, particles with an irregular shape and higher aspect ratios have been found to have a greater ability in stabilizing emulsions and foams (and at lower concentrations) compared to synthetic particles of spherical shape [19].

A special type of particles that display some similarities to surfactants and polymers are known as Janus particles. These are amphiphilic particles, composed of two or more regions with distinct physicochemical properties, that can self-assemble in bulk media and readily adsorb to fluid interfaces, remarkably lowering the interfacial tension; for this reason, they are also called "colloidal surfactants" [14, 23, 24]. They can be synthesized in geometrically different shapes and chemical compositions with high uniformity and precision [14]. Polysaccharides, such as, alginates, chitosan, pectin, cellulose and heparin, have been used to produce bio-based Janus particles [25, 26].

Another interesting type of emulsifying particles are microgels, which are soft deformable gel-like particles made up of aggregated or cross-linked polymer networks [22]. These microgels can swell in aqueous solvents and rearrange at the oil–water interface, resulting in thick and mechanically resilient layers. Owing to the amphiphilic character of their polymeric constituents, most microgels are inherently surface active at oil–water and air-water interfaces and, as rigid particles, they also irreversibly adsorb at the interfaces [22]. Synthetic microgels offer an additional feature that arises as a direct consequence of their combined polymeric and particulate character. They have the potential to effectively stabilize water-in-water emulsions, which are mixed solutions of thermodynamically incompatible polymers, producing two immiscible aqueous phases, and where the effective thickness of the interface is defined on a length scale considerably greater than the molecular dimensions of a conventional emulsifier [22, 27]. However, microgels based on physical cross-linking of biopolymers are rather novel and much of their behavior at interfaces remains unclear [15]. Two examples of natural ingredients that exhibit microgel-like characteristics are casein micelles (in their native form) and whey proteins and gelatinized starch granules (upon heat treatment). However, in order to mimic the special features of the synthetic microgels, these traditional food-grade microgels need more pronounced long-term structural stability under conventional processing and storage conditions, which typically requires the introduction of additional covalent cross-links within the aggregated biopolymer-based entity [22].

3. Cellulose: a versatile source of emulsifiers

3.1 Physicochemical characteristics of cellulose and the various morphological forms

Cellulose is a polysaccharide composed of glucose monomers, the anhydroglucose units (AGUs), linked by β -(1–4) glycosidic bonds. These β-linked AGUs adopt the 4C1 chain conformation, which is the conformation with the lowest free energy of the molecule. Consequently, the three polar hydroxyl groups in each AGU are located on the equatorial positions of the rings, and the hydrogen atoms of the non-polar C–H bonds are located on the axial positions [1]. This structural anisotropy is what gives cellulose its amphiphilic nature [28]. Due to the large number of hydroxyl groups within a cellulose molecule, both intra- and intermolecular hydrogen bonding occur and various types of supramolecular semi-crystalline structures can be formed. It is believed that intramolecular hydrogen bonding is responsible for the single-chain conformation and stiffness, while the intermolecular hydrogen bonding would be responsible for the sheet-like arrangement of the native polymer [1]. However, the stacking of these sheets into the three-dimensional crystalline supramolecular structures must involve hydrophobic interactions, as it was shown trough molecular dynamic simulations, and moreover was observed, many years ago, in native cellulose biosynthesis [1, 29, 30].

Hydrophobic interactions between cellulose molecules make, in combination with favorable packing conditions (and thus a low energy) in the crystalline state, cellulose insoluble in water [31]. Solubility can be achieved by ionizing cellulose, which occurs at extreme pH's. Solubility in water can also be aided by addition of co-solutes that weaken hydrophobic interactions. Derivatization of cellulose is also found to generally enhance solubility strongly, which can be referred to packing constraints in the solid state. Thus, to make cellulose soluble in aqueous solutions, the crystalline packing has to be disrupted, and this can be achieved, for example, by chemical modifications via etherification reactions in alkaline media, resulting in water-soluble cellulose ethers [32, 33]. These cellulose derivatives keep the amphiphilic properties of cellulose, as can be seen from their association with surfactants and their adsorption at the air-water and oil–water interfaces [34–39].

A fairly simple way of converting cellulose into a versatile class of new materials is through a dissolution-regeneration process. The regeneration of cellulose occurs when a coagulant ("anti-solvent") gets in contact with a cellulose solution or dope, leading to a solvent exchange and subsequent aggregation of the cellulose chains. The organization of the molecules in the regenerated materials (e.g., fibers, films, foams, particles) and their properties are strongly influenced by the dissolution state of cellulose (molecular cellulose, partially dissolved, crystalline or amorphous aggregates), as a well as the nature of the coagulant used [40, 41].

Cellulose can also be shaped into micro- and nanoparticles of different colloidal structure. Acid or mechanical treatments are usually applied to deconstruct the cellulose fibers into crystalline or semi-crystalline nanocelluloses [20, 42–44]. Partial decomposition of cellulose fibers, by acid treatment and cellulase-catalyzed hydrolysis, yields powdery microcrystalline cellulose (MCC), such as commercial Avicel®, with DP values between 150 and 300 [45]. Avicel® still contains both amorphous and crystalline portions. On the other hand, nanocrystalline cellulose (NCC) is obtained by strong acid hydrolysis. During the chemical process, the more readily accessible amorphous regions are completely disrupted deliberating rod-like crystal sections, whose sizes are dependent on the time and temperature of the reaction. The dimensions of the isolated NCC are also found to be strongly influenced by the degree of crystallinity of cellulose, which, in turn, is dependent on the natural source. Cotton, wood and Avicel® usually yield highly crystalline nanorods (90% crystallinity) with a narrow distribution of sizes (5–10 nm in width and 100–300 nm in length), whereas sources, such as tunicin (extracted from the tunicates), bacteria and algae, yield crystals with higher polydispersity and larger dimensions [42]. NCC forms stable suspensions in water by application of a mechanical force, typically sonication. Its surface properties are determined by the mineral acid and the reaction conditions used during its extraction. NCC prepared with hydrochloric acid (HCl) is weakly negatively charged, while it exhibits a strong repulsive character if prepared with sulfuric acid (H_2SO_4), since approximately one tenth of the glucose units is functionalized with the anionic sulfate ester groups. Thus, NCC prepared with H_2SO_4 give suspensions with higher colloidal stability than NCC prepared with HCl.

Micro- and nanofibrillated (MFC/NFC) celluloses are obtained by extruding wood pulp suspensions trough mechanical devices (high-pressure homogenizers), which results in fiber delamination and deliberation of the fibrils, usually being tens of nanometers wide and lengths ranging from several nanometers to several micrometers (i.e., 5–60 nm in width and 100 nm to several micrometers in length) [42]. This type of nanocelluloses are usually less crystalline than NCC, since they still own part of the amorphous domains, and have higher aspect ratios [5, 46]. In aqueous solutions, the fiber-like morphology and high aspect ratio, typically drive gel-like behavior due to entanglements between the microfibrils.

3.2 Nanocelluloses at interfaces

Several researchers have demonstrated the ability of cellulose particles to self-assemble at oil–water interfaces and to stabilize o/w emulsions without the aid of classical surfactants [47–49]. It is believed that the amphiphilic character of nanocellulose resides in its crystalline organization at the elementary "brick" level, and thus, cellulose nanocrystals have both hydrophilic and hydrophobic edges that are preferentially wetted by water and oil phases, respectively [49]. The wettability properties of cellulose particles may be tuned by surface hydrophobization, due to the presence of many reactive hydroxyl groups, and w/o emulsions can be formed [50–52]. Most of the particles from biological origins, such as cellulose, chitosan, or starch, show an irregular shape and are polydisperse in size and morphology. However, this structural anisotropy may be very beneficial for emulsion formation and stability. Particles with high aspect ratios are capable of stabilizing biphasic systems at lower concentrations compared to systems containing spherical particles [53]. Particles with such a well-defined shape are usually derived from inorganic materials, like silica and these have been extensively studied because of their availability in different sizes with narrow size distributions and chemical surface tunability. However, their lack of biocompatibility and biodegradability restricts their use in food and pharmaceutical applications [54]. For this reason, the study and characterization of materials from biological origins have gained increasing attention, and many efforts have been made in the food and pharmaceutical industries in order to develop new food-grade particles [19, 55]. It was early noticed that MCC particles have the ability to stabilize conventional o/w emulsions, and multiple emulsions systems of w/o/w type, with the aid of a hydrophobic surfactant for the stabilization of the internal w/o interface [56, 57]. These MCC particles form a network around the emulsified oil droplets that provides a mechanical barrier against coalescence, and, beyond that, the non-adsorbed particles may act as thickener agents in the continuous aqueous phase. MCC particles have also the ability to reduce lipid oxidation, one of the major concerns among food manufacturers due to its negative effects on food quality [55]. More recently, nanocelluloses, such as MFC/NFC and NCC, have been increasingly in focus for having a better performance than MCC, owing their smaller sizes and more regular shapes [58]. NCCs with low aspect ratios (shorter) have a dense organization at the interface and cover better the oil surface, while NCCs with high aspect ratios (longer) typically form a network around the droplet with relatively low coverage. Therefore, shorter NCCs have better emulsification efficiency and long-term stability, since higher droplet coverage usually means smaller droplet size [59, 60]. On the other hand, long nanofibrils (NFC) with a high aspect ratio also tend to form bigger droplets resultant from a lower surface coverage, but the fibers protrude in the continuous phase forming a strong network that is able to physically hinder droplet coalescence [59]. As mentioned, the colloidal stability of NCC is controlled by their surface charge resulting from the acid hydrolysis with various acids (e.g., H_2SO_4 or HCl). The higher the charge density the better their colloidal stability, but their ability to efficiently stabilize emulsions is reduced. Thus, the anionic charges on the surface of the nanocrystals control their tendency to be dispersed in water in relation to being adsorbed at the oil–water interface and, therefore, the particle polarity must be confined to a limited range. A surface charge density lower than ca. 0.03 e/nm^2 is ideal for the effectiveness of NCC as an emulsifier and stabilizer, usually achieved by HCl hydrolysis. NCC with sulfate groups, resultant from the hydrolysis with H_2SO_4, possess a high surface charge density (e.g., 0.123 e/nm^2), and the charges may undergo desulfation or may be screened by salt addition, to tune their amphiphilicity [49, 61]. Nanocellulose-stabilized emulsions are generally thermally stable, but in the presence of charges their stability against pH and ionic strength may decrease [58, 62]. NCC are able to form stable o/w high internal phase emulsions (HIPEs) containing volume fractions of oil as high as 0.9, at very low NCC concentrations (< 0.1 wt.%) [61]. Hydrophobized nanocellulose has been also explored to form w/o HIPEs [51]. Double emulsions of both o/w/o and w/o/w have been prepared by using a combination of native and hydrophobized NCC and NFC [63, 64]. Apart from the outstanding physical stability against coalescence, nanocelluloses also afford oxidative stability and lipid digestion control due to the dense interfacial layer formed [60].

3.3 Cellulose derivatives at interfaces

Cellulose derivatives produced by etherification reactions are generally water-soluble and surface-active. Therefore, cellulose ethers are a major class of commercially important water-soluble polymers, from construction products, ceramics, and paints to foods, cosmetics, and pharmaceuticals [32, 33, 65, 66]. Cellulose ethers are commonly made by reacting alkali cellulose with the appropriate reagents to substitute the hydroxyl groups of the AGU monomers by either alkyl, hydroxyalkyl or carboxyalkyl groups [66]. Methyl cellulose (MC), ethyl cellulose (EC), hydroxyethyl cellulose (HEC), hydroxypropyl cellulose (HPC), and their derivatives, are common products of those reactions. Their functionality and solubility in water depend on the type of substituent, degree and pattern of substitution, and molecular weight [33, 67]. The non-ionic cellulose ethers, such as, MC, HPMC and EHEC and their hydrophobically modified versions, have been mostly used to produce o/w emulsions due to their water-solubility [68–71], but EC can be used to stabilize w/o emulsions [72]; this change-over from MC to EC illustrates the subtle role of polarity and illustrates the applicability of Bancroft's rule. The emulsion stabilization due to cellulose ethers is the result of the combined effects of: a) reduction of the interfacial tension, arising from the balance between polar and non-polar groups; b) adsorption of thick layers, forming a physical barrier with strong steric repulsion; and c) the viscosity increase of the continuous phase, constraining droplet dynamics [7, 71, 73]. Cellulose derivatives often show a dual effect, as stabilizing and emulsifying agents [73]. However, carboxymethyl cellulose (CMC) which is an anionic polymer, cannot efficiently stabilize emulsions by itself due to its highly polar character. However, it can assist emulsion stabilization by controlling the viscosity of the continuous medium [5, 16]. In general, cellulose ethers provide good stability against droplets aggregation due to the strong steric repulsions between the adsorbed polymer layers of two approaching droplets, and due to the increased viscosity of the systems. One of the main advantages of using non-ionic polysaccharides, such as the cellulose ethers in this specific case, is their high stability against environmental stresses, such as, pH, ionic strength, and temperature. This is particularly important in food and pharmaceutical applications, where complex environments are encountered. Cellulose ethers also provide good oxidative stability to the core materials and delay lipid digestion of o/w emulsions, provided that the physical barrier and thickened aqueous phase slows down the diffusion of pro-oxidants and lipases [71, 74]. Lipid digestion is even further reduced using cellulose ethers when compared to calcium-caseinate, a common food emulsifier. Additionally, the thermo-gelling ability of cellulose ethers, in particular HMC, makes it possible to obtain emulsions with high consistency during gastric digestion, contributing to slow down the gastric digestion and increase fullness and satiety perceptions [71].

3.4 Molecular and regenerated cellulose at interfaces

The behavior of molecularly dissolved cellulose at interfaces is expected to resemble that of typical cellulose derivatives or any semiflexible amphiphilic polymer that shows interfacial activity, i.e., the tendency to adsorb at oil–water interfaces and reduce the free energy between the two phases. However, due to its dissolution limitations, the properties of molecularly dissolved cellulose and its potential in emulsions formation and stability are clearly much less explored. Nevertheless, recent studies have confirmed the stated hypothesis. Molecular dynamics simulations indicate that molecularly dispersed cellulose gradually assembles at the oil–water interface eventually surrounding the oil droplet [75]. Experimentally, molecular cellulose dissolved in H_3PO_4 (aq.) was found to adsorb at the oil–water interface and decrease the interfacial tension (IFT) between the two phases (**Figure 2**).

The decrease in IFT is similar in magnitude to that of the non-ionic cellulose derivatives MC and HPMC, for the same polymer concentration (0.1 wt.%) and same type of oil (liquid paraffin) [34]. Yet, cellulose-stabilized emulsions formed in H_3PO_4 (aq.) were found to be short-lived, as oil was separating from the emulsions and floating to the top within 24 h. However, by subsequently adding water to the dispersions during emulsification, the properties of the emulsions changed

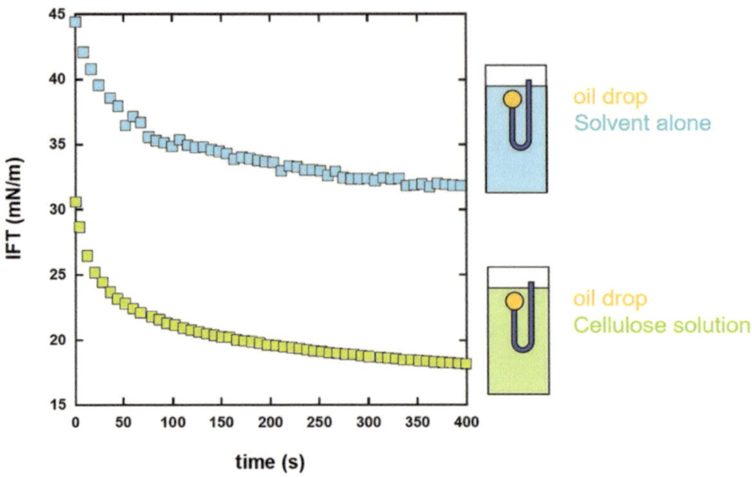

Figure 2.
Effect of dissolved cellulose on the interfacial tension oil-aqueous medium.

dramatically, and there was no evidence of oil separation over one year of storage (**Figure 3**). This effect was attributed to a decrease in cellulose solvency in H_3PO_4 (aq.) by the addition of an anti-solvent (water), which promoted a greater affinity for the oil–water interface, leading to the outstanding stability against macroscopic phase separation of the oil.

There are two ways of using native cellulose to stabilize o/w emulsions without the need of further modifications. One, is by following the dissolution-regeneration-emulsification approach, resulting in Pickering emulsions of solid or soft cellulose particles (microgels), since the oil is either dispersed in a water suspension of cellulose particles or in a water suspension of cellulose microgels, respectively [76–83].

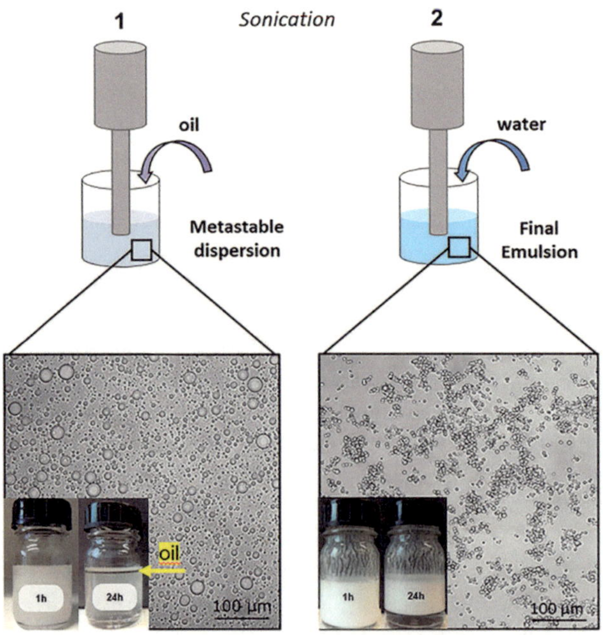

Figure 3.
Two step emulsification procedure for cellulose dissolved in phosphoric acid solvent.

Another way, is to follow the dissolution-emulsification-("in situ")regeneration approach, where the oil is directly dispersed in the cellulose solution, and regeneration takes place at the oil–water interface ("in situ") [34, 84]. This way, the oil droplets seem to be stabilized by a "cellulose film" with a smooth appearance, contrasting with the rough networks and particulate appearances of Pickering emulsions (**Figure 4**) [7, 84]. Another fundamental difference between the two described approaches is related to the existence of dissolved cellulose during the oil emulsification, which acts as a polymeric surfactant by decreasing the IFT, and possibly contributing for a reduction in droplet size. Overall, the emulsions produced by both methods display very good stability against droplet coalescence which can be referred to the irreversible adsorption of cellulose onto the droplet surfaces [34, 78, 79, 85]. Soft cellulose microgels also impart an outstanding stability against flocculation because of the thick viscoelastic layers formed at the interface [15]. The mechanism behind droplet stabilization in emulsions prepared with cellulose particles is similar to that operating for nanocelluloses as described above, i.e., a combination of Pickering adsorption and network stabilization, often showing gel-like characteristics upon a concentration increase of cellulose [79, 82–84].

Moreover, the resulting emulsions are remarkably stable against environmental changes, such as, pH, ionic strength, and temperature, which makes them good candidates for target delivering [80, 82, 86]. Cellulose regenerated particles have also been shown to improve the physical stability of emulsions stabilized by sodium-caseinate, a milk-protein commonly used as a food emulsifier, promoting adsorption of the protein and thickening the continuous phase [87]. Very little has been done regarding the stabilization of w/o emulsions since cellulose particles are better wetted by water than oil. However, it has been suggested that the presence of a water–oil interface when regenerating cellulose affects the conformation of the cellulose molecules and so the way they reassemble. Therefore w/o emulsions are possible to form, but they have poorer stability compared to o/w emulsions [78]. More recently, a "hydrophobic" cellulose microgel was developed to stabilize w/o emulsions, by coagulating cellulose in the presence of a coagulant and sunflower

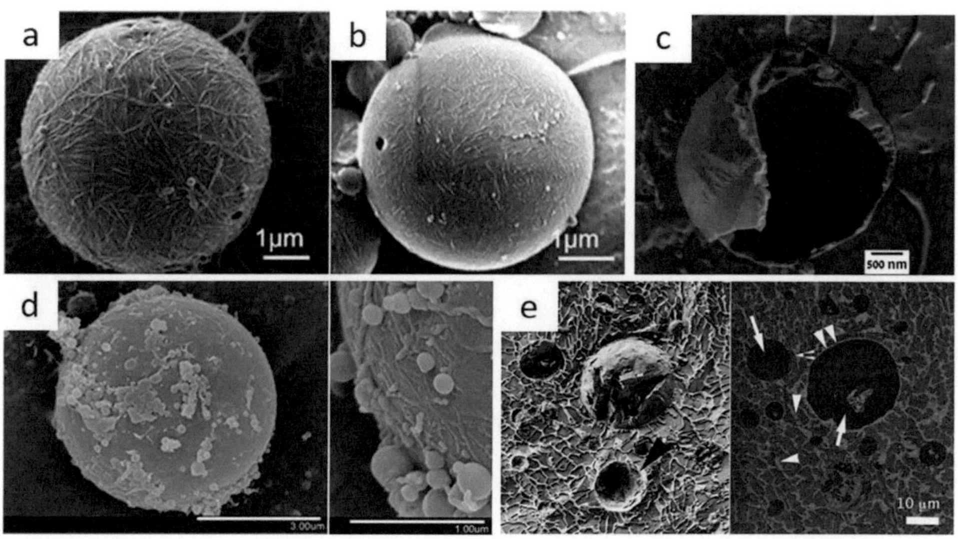

Figure 4.
Different morphologies of cellulose-stabilized emulsions. Particle-stabilized emulsions by longer NCC (bacterial cellulose) (a), shorter NCC (b), and regenerated cellulose (dissolution-regeneration-emulsification approach) (d). Emulsions prepared from molecular solutions of cellulose (dissolution-emulsification-"in-situ" regeneration approach) (c and e). Reprinted (adapted) with permission from Ref. [49, 77, 80, 84].

oil [88]. The resultant microgels were more easily dispersed in oil than water, and stable emulsions w/o emulsions were formed. The simplicity and versatility of the dissolution-regeneration approaches open many new possibilities for the functionalization of cellulose and its applicability in both o/w and w/o emulsions.

4. Conclusions

The list of emulsion formulations having a remarkable impact in our lives is vast, and therefore, it is not surprising that natural molecules have been emerging as important players to partially or completely replace the available non-sustainable options. The future leading role of cellulose as an effective stabilizing agent is unquestionable and opens a new era of sustainable, biocompatible, and value-added functional materials. Surfactant-free emulsions have recently been developed using all forms of cellulose (crystalline, fibrillated, molecular and regenerated), providing a strong support for the vision of cellulose as an amphiphilic molecule, capable of acting as a polymeric surfactant and a Pickering stabilizer. Structural differences and mechanisms of emulsion stabilization between the different cellulose forms have been presented in this chapter. In general, and given the right conditions, cellulose coatings are a powerful mechanical barrier against coalescence, lipid oxidation and lipid digestion. Non-adsorbed cellulose forms a 3D network in the continuous phase, that constrains droplet movements and enhances kinetic stability. The colloidal assembly of cellulose particles when liquid interfaces of notably different polarities are present might serve as a template for the synthesis of new functional microcapsules. The dissolution-regeneration process is highlighted as an important approach of making cellulose-based emulsions, whose hydrophilic–lipophilic balance can be simply tuned by playing with solvent quality and regeneration coagulant(s). Furthermore, the exceptional stability against environmental stresses (pH, ionic strength and temperature) makes the cellulose regenerated coatings potential candidates for target delivery in complex conditions.

Acknowledgements

This research was funded by the Swedish Research Council (Vetenskapsrådet), grant number 2015-04290. Bruno Medronho acknowledges the financial support from FCT via the projects PTDC/ASP-SIL/30619/2017, UIDB/05183/2020, and the researcher grant IF/01005/2014. Björn Lindman acknowledges the support from the Coimbra Chemistry Centre (CQC), funded by FCT through the Project UID/QUI/00313/2020. The authors also acknowledge the support from Treesearch.se.

Conflict of interest

The authors declare no conflict of interest.

Author details

Carolina Costa[1*], Bruno Medronho[2], Björn Lindman[3,4], Håkan Edlund[1] and Magnus Norgren[1]

1 FSCN, Surface and Colloid Engineering, Mid Sweden University, Sundsvall, Sweden

2 Faculty of Sciences and Technology, MED (Mediterranean Institute for Agriculture), Environment and Development, University of Algarve, Faro, Portugal

3 Physical Chemistry, University of Lund, Lund, Sweden

4 Coimbra Chemistry Center (CQC), Department of Chemistry, University of Coimbra, Coimbra, Portugal

*Address all correspondence to: carolina.costa@miun.se

References

[1] Klemm, D., H.-P. Schmauder, and T. Heinze, *Cellulose*, in *Biopolymers Online*. 2005, Wiley-VCH Verlag GmbH & Co. KGaA.

[2] Bajpai, P., *Chapter 12 - Pulping Fundamentals*, in *Biermann's Handbook of Pulp and Paper (Third Edition)*, P. Bajpai, Editor. 2018, Elsevier. p. 295-351.

[3] Radotić, K. and M. Mićić, *Methods for Extraction and Purification of Lignin and Cellulose from Plant Tissues*, in *Sample Preparation Techniques for Soil, Plant, and Animal Samples*, M. Micic, Editor. 2016, Springer New York: New York, NY. p. 365-376.

[4] Szymańska-Chargot, M., et al., *Isolation and Characterization of Cellulose from Different Fruit and Vegetable Pomaces.* Polymers, 2017. **9**(10).

[5] He, X., et al., *Cellulose and cellulose derivatives: Different colloidal states and food-related applications.* Carbohydrate Polymers, 2021. **255**: p. 117334.

[6] Walstra, P., *8 - Emulsions*, in *Fundamentals of Interface and Colloid Science*, J. Lyklema, Editor. 2005, Academic Press. p. 8.1-8.94.

[7] Costa, C., et al., *Emulsion Formation and Stabilization by Biomolecules: The Leading Role of Cellulose.* Polymers, 2019. **11**(10).

[8] Dickinson, E., *Hydrocolloids at interfaces and the influence on the properties of dispersed systems.* Food Hydrocolloids, 2003. **17**(1): p. 25-39.

[9] Bouyer, E., et al., *Proteins, polysaccharides, and their complexes used as stabilizers for emulsions: Alternatives to synthetic surfactants in the pharmaceutical field?* International Journal of Pharmaceutics, 2012. **436**(1-2): p. 359-378.

[10] McClements, D.J. and C.E. Gumus, *Natural emulsifiers - Biosurfactants, phospholipids, biopolymers, and colloidal particles: Molecular and physicochemical basis of functional performance.* Advances in Colloid and Interface Science, 2016. **234**: p. 3-26.

[11] Dickinson, E., *Hydrocolloids as emulsifiers and emulsion stabilizers.* Food Hydrocolloids, 2009. **23**(6): p. 1473-1482.

[12] Holmberg, K., et al., *Emulsions and Emulsifiers*, in *Surfactants and Polymers in Aqueous Solution*. 2002, John Wiley & Sons, Ltd. p. 451-471.

[13] Dickinson, E., *Food emulsions and foams: Stabilization by particles.* Current Opinion in Colloid & Interface Science, 2010. **15**(1-2): p. 40-49.

[14] Kumar, A., et al., *Amphiphilic Janus particles at fluid interfaces.* Soft Matter, 2013. **9**(29): p. 6604-6617.

[15] Lefroy, K.S., B.S. Murray, and M.E. Ries, *Advances in the use of microgels as emulsion stabilisers and as a strategy for cellulose functionalisation.* Cellulose, 2021. **28**(2): p. 647-670.

[16] Mikkonen, K.S., *Strategies for structuring diverse emulsion systems by using wood lignocellulose-derived stabilizers.* Green Chemistry, 2020. **22**(4): p. 1019-1037.

[17] Bai, L., et al., *Adsorption and Assembly of Cellulosic and Lignin Colloids at Oil/Water Interfaces.* Langmuir, 2019. **35**(3): p. 571-588.

[18] Dickinson, E., *Biopolymer-based particles as stabilizing agents for emulsions and foams.* Food Hydrocolloids, 2017. **68**: p. 219-231.

[19] Lam, S., K.P. Velikov, and O.D. Velev, *Pickering stabilization of foams and*

emulsions with particles of biological origin. Current Opinion in Colloid & Interface Science, 2014. **19**(5): p. 490-500.

[20] Salas, C., et al., *Nanocellulose properties and applications in colloids and interfaces.* Current Opinion in Colloid & Interface Science, 2014. **19**(5): p. 383-396.

[21] Tavernier, I., et al., *Food-grade particles for emulsion stabilization.* Trends in Food Science & Technology, 2016. **50**: p. 159-174.

[22] Dickinson, E., *Exploring the frontiers of colloidal behaviour where polymers and particles meet.* Food Hydrocolloids, 2016. **52**: p. 497-509.

[23] Binks, B.P. and P.D.I. Fletcher, *Particles Adsorbed at the Oil–Water Interface: A Theoretical Comparison between Spheres of Uniform Wettability and "Janus" Particles.* Langmuir, 2001. **17**(16): p. 4708-4710.

[24] Wang, Z. and Y. Wang, *Tuning Amphiphilicity of Particles for Controllable Pickering Emulsion.* Materials, 2016. **9**(11).

[25] Marschelke, C., A. Fery, and A. Synytska, *Janus particles: from concepts to environmentally friendly materials and sustainable applications.* Colloid and Polymer Science, 2020. **298**(7): p. 841-865.

[26] Sheikhi, A. and T.G.M. van de Ven, *Colloidal aspects of Janus-like hairy cellulose nanocrystalloids.* Current Opinion in Colloid & Interface Science, 2017. **29**: p. 21-31.

[27] Esquena, J., *Water-in-water (W/W) emulsions.* Current Opinion in Colloid & Interface Science, 2016. **25**: p. 109-119.

[28] Yamane, C., et al., *Two Different Surface Properties of Regenerated Cellulose due to Structural Anisotropy.* Polym. J, 2006. **38**(8): p. 819-826.

[29] Bergenstråhle, M., et al., *Simulation studies of the insolubility of cellulose.* Carbohydrate Research, 2010. **345**(14): p. 2060-2066.

[30] Cousins, S.K. and R.M. Brown, *Cellulose I microfibril assembly: computational molecular mechanics energy analysis favours bonding by van der Waals forces as the initial step in crystallization.* Polymer, 1995. **36**(20): p. 3885-3888.

[31] Lindman, B., et al., *The relevance of structural features of cellulose and its interactions to dissolution, regeneration, gelation and plasticization phenomena.* Physical Chemistry Chemical Physics, 2017. **19**(35): p. 23704-23718.

[32] Jedvert, K. and T. Heinze, *Cellulose modification and shaping – a review.* Journal of Polymer Engineering, 2017. **37**(9): p. 845-860.

[33] Seddiqi, H., et al., *Cellulose and its derivatives: towards biomedical applications.* Cellulose, 2021. **28**(4): p. 1893-1931.

[34] Costa, C., et al., *Interfacial activity and emulsion stabilization of dissolved cellulose.* Journal of Molecular Liquids, 2019: p. 111325.

[35] Carlsson, A., G. Karlstroem, and B. Lindman, *Characterization of the interaction between a nonionic polymer and a cationic surfactant by the Fourier transform NMR self-diffusion technique.* The Journal of Physical Chemistry, 1989. **93**(9): p. 3673-3677.

[36] Kamenka, N., et al., *Electrical Conductivity, Self-Diffusion, and Fluorescence Probe Investigations of the Interaction between Sodium Dodecyl Sulfate and Ethyl(hydroxyethyl)cellulose.* The Journal of Physical Chemistry, 1994. **98**(27): p. 6785-6789.

[37] Persson, B., S. Nilsson, and R. Bergman, *Dynamic Surface Tension of*

Dilute Aqueous Solutions of Nonionic Cellulose Derivatives in Relation to Other Macromolecular Characterization Parameters. Journal of Colloid and Interface Science, 1999. **218**(2): p. 433-441.

[38] Persson, B., S. Nilsson, and L.-O. Sundelöf, *On the characterization principles of some technically important water-soluble nonionic cellulose derivatives. Part II: Surface tension and interaction with a surfactant.* Carbohydrate Polymers, 1996. **29**(2): p. 119-127.

[39] Nasatto, P.L., et al., *Interfacial Properties of Methylcelluloses: The Influence of Molar Mass.* Polymers, 2014. **6**(12).

[40] Alves, L., et al., *Dissolution state of cellulose in aqueous systems. 1. Alkaline solvents.* Cellulose, 2016. **23**(1): p. 247-258.

[41] Alves, L., et al., *Dissolution state of cellulose in aqueous systems. 2. Acidic solvents.* Carbohydrate Polymers, 2016. **151**: p. 707-715.

[42] Klemm, D., et al., *Nanocelluloses: A New Family of Nature-Based Materials.* Angewandte Chemie International Edition, 2011. **50**(24): p. 5438-5466.

[43] Thomas, B., et al., *Nanocellulose, a Versatile Green Platform: From Biosources to Materials and Their Applications.* Chemical Reviews, 2018. **118**(24): p. 11575-11625.

[44] Trache, D., et al., *Microcrystalline cellulose: Isolation, characterization and bio-composites application—A review.* International Journal of Biological Macromolecules, 2016. **93**: p. 789-804.

[45] Klemm, D., et al., *Cellulose: Fascinating Biopolymer and Sustainable Raw Material.* Angewandte Chemie International Edition, 2005. **44**(22): p. 3358-3393.

[46] Tingaut, P., T. Zimmermann, and G. Sebe, *Cellulose nanocrystals and microfibrillated cellulose as building blocks for the design of hierarchical functional materials.* Journal of Materials Chemistry, 2012. **22**(38): p. 20105-20111.

[47] Kalashnikova, I., et al., *Cellulosic nanorods of various aspect ratios for oil in water Pickering emulsions.* Soft Matter, 2013. **9**(3): p. 952-959.

[48] Kalashnikova, I., et al., *New Pickering Emulsions Stabilized by Bacterial Cellulose Nanocrystals.* Langmuir, 2011. **27**(12): p. 7471-7479.

[49] Kalashnikova, I., et al., *Modulation of Cellulose Nanocrystals Amphiphilic Properties to Stabilize Oil/Water Interface.* Biomacromolecules, 2012. **13**(1): p. 267-275.

[50] Andresen, M. and P. Stenius, *Water-in-oil Emulsions Stabilized by Hydrophobized Microfibrillated Cellulose.* Journal of Dispersion Science and Technology, 2007. **28**(6): p. 837-844.

[51] Lee, K.-Y., et al., *Phase Behavior of Medium and High Internal Phase Water-in-Oil Emulsions Stabilized Solely by Hydrophobized Bacterial Cellulose Nanofibrils.* Langmuir, 2014. **30**(2): p. 452-460.

[52] Xhanari, K., K. Syverud, and P. Stenius, *Emulsions Stabilized by Microfibrillated Cellulose: The Effect of Hydrophobization, Concentration and O/W Ratio.* Journal of Dispersion Science and Technology, 2011. **32**(3): p. 447-452.

[53] Madivala, B., et al., *Exploiting particle shape in solid stabilized emulsions.* Soft Matter, 2009. **5**(8): p. 1717-1727.

[54] Asefa, T. and Z. Tao, *Biocompatibility of Mesoporous Silica Nanoparticles.* Chemical Research in Toxicology, 2012. **25**(11): p. 2265-2284.

[55] Kargar, M., et al., *Investigation into the potential ability of Pickering emulsions (food-grade particles) to enhance the oxidative stability of oil-in-water emulsions.* Journal of Colloid and Interface Science, 2012. **366**(1): p. 209-215.

[56] Oza, K.P. and S.G. Frank, *Microcrystalline celulose stabilized emulsions.* Journal of Dispersion Science and Technology, 1986. 7(5): p. 543-561.

[57] Oza, K.P. and S.G. Frank, *Multiple emulsions stabilized by colloidal microcrystalline cellulose* Journal of Dispersion Science and Technology, 1989. **10**(2): p. 163-185.

[58] Fujisawa, S., E. Togawa, and K. Kuroda, *Nanocellulose-stabilized Pickering emulsions and their applications.* Science and technology of advanced materials, 2017. **18**(1): p. 959-971.

[59] Capron, I., O.J. Rojas, and R. Bordes, *Behavior of nanocelluloses at interfaces.* Current Opinion in Colloid & Interface Science, 2017. **29**: p. 83-95.

[60] Dai, H., et al., *Recent advances on cellulose nanocrystals for Pickering emulsions: Development and challenge.* Trends in Food Science & Technology, 2020. **102**: p. 16-29.

[61] Capron, I. and B. Cathala, *Surfactant-Free High Internal Phase Emulsions Stabilized by Cellulose Nanocrystals.* Biomacromolecules, 2013. **14**(2): p. 291-296.

[62] Wen, C., et al., *Preparation and stabilization of d-limonene Pickering emulsions by cellulose nanocrystals.* Carbohydrate Polymers, 2014. **112**: p. 695-700.

[63] Cunha, A.G., et al., *Preparation of Double Pickering Emulsions Stabilized by Chemically Tailored Nanocelluloses.* Langmuir, 2014. **30**(31): p. 9327-9335.

[64] Carrillo, C.A., T.E. Nypelö, and O.J. Rojas, *Cellulose nanofibrils for one-step stabilization of multiple emulsions (W/O/W) based on soybean oil.* Journal of Colloid and Interface Science, 2015. **445**: p. 166-173.

[65] Hon, D.N.S., *Cellulose: Chemistry and Technology*, in *Encyclopedia of Materials: Science and Technology*, K.H.J. Buschow, et al., Editors. 2001, Elsevier: Oxford. p. 1039-1045.

[66] Murray, J.C.F., *25 - Cellulosics*, in *Handbook of Hydrocolloids (Second Edition)*, G.O. Phillips and P.A. Williams, Editors. 2009, Woodhead Publishing. p. 710-723.

[67] Akiyama, E., et al., *Thickening properties and emulsification mechanisms of new derivatives of polysaccharide in aqueous solution: 2. The effect of the substitution ratio of hydrophobic/hydrophilic moieties.* Journal of Colloid and Interface Science, 2007. **311**(2): p. 438-446.

[68] Nasatto, L.P., et al., *Methylcellulose, a Cellulose Derivative with Original Physical Properties and Extended Applications.* Polymers, 2015. 7(5).

[69] Schulz, M.B. and R. Daniels, *Hydroxypropylmethylcellulose (HPMC) as emulsifier for submicron emulsions: influence of molecular weight and substitution type on the droplet size after high-pressure homogenization.* European Journal of Pharmaceutics and Biopharmaceutics, 2000. **49**(3): p. 231-236.

[70] Davies, R.E.M. and J.M. Rowson, *Water-Soluble Cellulose Derivatives* Uses as Primary Emulsifyinga Gents. Part 1.* Journal of Pharmacy and Pharmacology, 1960. **12**(1): p. 154-162.

[71] Borreani, J., et al., *Oil-in-water emulsions stabilised by cellulose ethers: stability, structure and in vitro digestion.* Food & Function, 2017. **8**(4): p. 1547-1557.

[72] Melzer, E., J. Kreuter, and R. Daniels, *Ethylcellulose: a new type of emulsion stabilizer.* European Journal of Pharmaceutics and Biopharmaceutics, 2003. **56**(1): p. 23-27.

[73] Karlberg, M., K. Thuresson, and B. Lindman, *Hydrophobically modified ethyl(hydroxyethyl)cellulose as stabilizer and emulsifying agent in macroemulsions.* Colloids and Surfaces A: Physicochemical and Engineering Aspects, 2005. **262**(1): p. 158-167.

[74] Berton-Carabin, C.C., L. Sagis, and K. Schroën, *Formation, Structure, and Functionality of Interfacial Layers in Food Emulsions.* Annual Review of Food Science and Technology, 2018. **9**(1): p. 551-587.

[75] Miyamoto, H., et al., *Molecular dynamics simulation of cellulose-coated oil-in-water emulsions.* Cellulose, 2017. **24**(7): p. 2699-2711.

[76] Napso, S., et al., *Cellulose gel dispersion: From pure hydrogel suspensions to encapsulated oil-in-water emulsions.* Colloids and Surfaces B: Biointerfaces, 2016. **137**: p. 70-76.

[77] Napso, S., et al., *Structural Analysis of Cellulose-Coated Oil-in-Water Emulsions Fabricated from Molecular Solution.* Langmuir, 2018. **34**(30): p. 8857-8865.

[78] Rein, D.M., R. Khalfin, and Y. Cohen, *Cellulose as a novel amphiphilic coating for oil-in-water and water-in-oil dispersions.* Journal of Colloid and Interface Science, 2012. **386**(1): p. 456-463.

[79] Jia, X., et al., *Stabilizing oil-in-water emulsion with amorphous cellulose.* Food Hydrocolloids, 2015. **43**: p. 275-282.

[80] Shen, W., et al., *Stabilizing beverage emulsions by regenerated celluloses.* Lwt-Food Science and Technology, 2016. **72**: p. 292-301.

[81] Alfassi, G., D.M. Rein, and Y. Cohen, *Cellulose emulsions and their hydrolysis.* Journal of Chemical Technology & Biotechnology, 2019. **94**(1): p. 178-184.

[82] Li, Z., et al., *Stability mechanism of O/W Pickering emulsions stabilized with regenerated cellulose.* Carbohydrate Polymers, 2018. **181**: p. 224-233.

[83] Zhu, Y., et al., *Cellulose gel dispersions: fascinating green particles for the stabilization of oil/water Pickering emulsion.* Cellulose, 2017. **24**(1): p. 207-217.

[84] Costa, C., et al., *Cellulose-stabilized oil-in-water emulsions: Structural features, microrheology, and stability.* Carbohydrate Polymers, 2021. **252**: p. 117092.

[85] Jia, X.J., et al., *Preparation and Characterization of Cellulose Regenerated from Phosphoric Acid.* Journal of Agricultural and Food Chemistry, 2013. **61**(50): p. 12405-12414.

[86] Medronho, B., et al., *Microrheology of novel cellulose stabilized oil-in-water emulsions.* Journal of Colloid and Interface Science, 2018. **531**: p. 225-232.

[87] Hu, H.-y., et al., *Effects of regenerated cellulose on oil-in-water emulsions stabilized by sodium caseinate.* Food Hydrocolloids, 2016. **52**: p. 38-46.

[88] Lefroy, K.S., et al., *A natural, cellulose-based microgel for water-in-oil emulsions.* Food Hydrocolloids, 2021. **113**: p. 106408.